PATHS OF LIFE

Three Rarámuri (Tarahumara) matachines—members of a religious society devoted to the Virgin Mary—offer food during Holy Week in order to restore God's strength and ensure his favor. (Photograph by John P. Schaefer)

PATHS OF LIFE

American Indians of the Southwest and Northern Mexico

EDITED BY

THOMAS E. SHERIDAN

& NANCY J. PAREZO

THE UNIVERSITY OF ARIZONA PRESS ||| TUCSON

Publication of this book was made possible in part by the generosity of the Southwestern Foundation for Education and Historic Preservation; the Southwestern Mission Research Center; the Marshall Foundation; SWCA, Inc., Environmental Consultants; and the Southwest Center at the University of Arizona.

First printing 1996
The University of Arizona Press

01   00   99   98   97   96   6   5   4   3   2   1

Library of Congress Cataloging-in-Publication Data
Paths of life : American Indians of the Southwest and northern Mexico
    / edited by Thomas E. Sheridan and Nancy J. Parezo.
        p.   cm.
    Includes bibliographical references and index.
    ISBN 0-8165-1549-2 (cloth : alk. paper). —
    ISBN 0-8165-1466-6 (paper : alk. paper)
        1. Indians of North America—Southwest, New.   2. Indians of
    Mexico—Mexico, North.   I. Sheridan, Thomas E.   II. Parezo, Nancy J.
    E78.S7P38   1996   95-32531
    979'.00497—dc20   CIP

British Cataloguing-in-Publication Data
A catalogue record for this book is available from the British Library.

To the memory of
Ned Spicer and Bob Netting
Scholars, Teachers, and Friends

CONTENTS

List of Plates, xi

List of Figures, xiii

Foreword, xvii
  Raymond H. Thompson

Introduction, xxiii
  Thomas E. Sheridan and Nancy J. Parezo

The Diné (Navajos): Sheep Is Life, 3
  Nancy J. Parezo
    SIDEBARS
    Creation Story: The Gathering of the Clans, 5
      Nancy J. Parezo
    Navajo Weaving, 20
      Nancy J. Parezo
    Learning to Be Navajo, 24
      Nancy J. Parezo

The Yoemem (Yaquis): An Enduring People, 35
  Thomas E. Sheridan
    SIDEBARS
    The Yoremem (Mayos), 41
      Thomas E. Sheridan
    Yaqui and Mayo Pahkola/Pahkora Masks, 50
      Thomas M. Kolaz

The Indé (Western Apaches): The People of the Mountains, 61
  Bruce E. Hilpert
    SIDEBARS
    Apache Creation Story: The Things Legends Are Made Of, 64
      Bruce E. Hilpert

The Chiricahua Apaches, 70
Thomas E. Sheridan

The Havasupais, Hualapais, and Yavapais: The Great Creator
Has Given Us This Country, 91
Nancy J. Parezo
SIDEBARS
How to Weave a Basket, 96
Nancy J. Parezo
Lucille J. Watahomigie: Scholar and Educator, 111
Nancy J. Parezo

The O'odham (Pimas and Papagos): The World Would Burn
Without Rain, 115
Thomas E. Sheridan
SIDEBARS
The Wi:gida Ceremony, 124
Thomas E. Sheridan
Diet and Diabetes among the O'odham, 130
Thomas E. Sheridan

The Rarámuri (Tarahumaras): When We Walk in Circles, 141
Thomas E. Sheridan
SIDEBARS
The Rarámuri and the Leadville Trail 100, 144
Thomas E. Sheridan
The Railroad and the Tourists, 159
Thomas E. Sheridan

The Nỉngwỉ (Southern Paiutes): The People of the
Northwestern Frontier, 163
Catherine S. Fowler and Nancy J. Parezo
SIDEBARS
Southern Paiute Baskets, 170
Nancy J. Parezo
Marking the Traditional Landscape, 175
Nancy J. Parezo
"The Navajo Wedding Basket," 183
Nancy J. Parezo

The Comcáac (Seris): People of the Desert and Sea, 187
Thomas E. Sheridan

SIDEBARS

Seri Ironwood Carving, 204

Thomas E. Sheridan

Seri Santos, 206

Diane Dittemore

The Colorado River Yumans: Relations on the River, 213

Bruce E. Hilpert

SIDEBARS

Dreaming for Power, 215

Bruce E. Hilpert

The Cocopa Game of Peón, 233

Nancy J. Parezo

The Hopis: Hopivotskwani, the Hopi Path of Life, 237

Nancy J. Parezo

SIDEBARS

Emergence to the Fourth World, 240

Nancy J. Parezo

The Homol'ovi Ruins of Northeastern Arizona, 244

E. Charles Adams

Women's Roles: The Heart of Hopi Society, 248

Nancy J. Parezo and Bruce E. Hilpert

References and Suggested Readings, 267

Contributors, 281

Plate and Figure Credits, 285

Index, 291

PLATES

following page 60

1 Navajo shell and turquoise necklaces
2 Navajo woman's dress
3 Navajo squash-blossom necklaces
4 Navajo Ganado-style rug
5 Navajo Wide Ruins rug
6 Mural of the *sea ania* (the Yaqui flower world)
7 Mayo *pahkora* masks
8 Yaqui *pahkola* masks and disk rattle
9 *Apache Creation Story* painting
10 Western Apache coiled basketry
11 San Carlos Apache agave wood fiddle
12 Western Apache bilobe canteen
13 Western Apache painted twined burden basket
14 Western Apache beaded buckskin moccasins
15 Western Apache war shield
16 Canyon Creek in the Grand Canyon in the late 1850s
17 Havasupai basketry bowl and jar
18 Hualapai twined jar and "fruit basket"
19 Two Yavapai coiled baskets
20 Hualapai beadwork

following page 156

21 Tohono O'odham *wi:gida* figures
22 Tohono O'odham women picking saguaro fruit in 1854
23 Saguaro fruit pickers
24 Tohono O'odham saguaro wine-making paraphernalia

25 O'odham bowl baskets

26 Akimel O'odham women in 1854

27 O'odham pottery

28 Tarahumara cross and rosary

29 Tarahumara violin

30 Tarahumara Easter objects

31 Tarahumara Fariseo dancers on Holy Thursday

32 A Tarahumara Soldado and Fariseo

33 Tarahumaras dance around a straw Judas

34 Tarahumara Judas figure

35 Southern Paiute basketry water jars

36 San Juan Paiute baskets

37 Southern Paiute wedding baskets

following page 236

38 Seri ceramics and figurines

39 Seri seaweed dolls

40 Seri carrying pouches

41 Seri necklaces

42 Seri coiled basketry

43 Seri santos

44 Mohave figures

45 Mohave warriors in the late 1850s

46 A Quechan ferry on the Colorado River in about 1854

47 Colorado River Indian Tribe arts and crafts

48 Colorado River Yuman tourist crafts

49 Maricopa pottery

50 Hopi art with Kachina motifs

51 Hopi belt and brocaded sash

52 A Hopi *tableta*

53 Hano polychrome jar

54 Hopi tiles

55 Contemporary Hopi jar

56 Hopi Niman kachina *tihu*

57 Clark Tenakhongva carving a *tihu*

FIGURES

Three Rarámuri (Tarahumara) matachines, *Frontispiece*, ii

Comcáac (Seri) women with usable parts of a butchered turtle, xxv

A Yoeme (Yaqui) deer dancer and three deer singers, xxvii

Mission San Xavier del Bac in 1863, xxx

Hualapai students at a BIA day school in Kingman, about 1896, xxi

A Diné (Navajo) woman with sheep in a pen, about 1950, xxxii

A Navajo shepherd and his flock at Pine Springs, near Chinle, in 1970, 4

Soldiers count Navajo prisoners at Fort Sumner in 1864 or 1866, 13

The treaty of 1868 between the United States and the Navajos, 15

Sheep being dipped after shearing, 16

Men prepare wool for market, about 1935, 17

A Navajo silversmith, about 1910, 18

The Hubbell Trading Post on the Navajo Reservation, 19

Navajo women weaving at their camp in Keams Canyon in 1893, 21

Brian Francis teasing an old ram at his mother's family's home in 1974, 24

Laura Ann Nez weaving a rug, 27

Navajo workers at the Fairchild Electronics Company plant, 1968, 31

Three Yaqui men at the boundary between Potam and Vicam, early 1940s, 37

Cajeme, the leader of the Yaqui resistance against the Mexican government in the 1880s, 42

Yaqui soldiers in about 1890, 45

Yaqui prisoners in about 1900, 46

Miguel Romero carrying a statue of San José from the Potam church, early 1940s, 54

Yaqui drummer Juan Tampaleo and his grandchildren, 1989, 57

Dutchy, Alchesay, and General George Crook in 1885, 67

Ration lines at the San Carlos Agency in about 1880, 69

A Western Apache (San Carlos) woman preparing mescal, about 1930, 74

An Apache family moving camp in about 1900, 75

A Western Apache cornfield in 1936, 78

A woman harvesting corn at East Fork in about 1900, 80

A Yavapai woman making a basket in about 1900, 96

A Yavapai family at their camp in about 1887, 95

A Havasupai settlement, 99

Qua-tha, a Yavapai scout in about 1882, 109

A Tohono O'odham calendar stick, 121

A Tohono O'odham woman's carrying basket, 123

A Tohono O'odham adobe winter home in 1894, 127

A Tohono O'odham field with a summer field house, 1975, 128

Akimel O'odham and Maricopas trading with Anglo American soldiers in 1846, 132

Men with horse teams building the Florence canal in 1887, 135

Coolidge Dam along the Gila River, 137

Rarámuri in front of thatched houses and stone storehouses, early 1890s, 146

Tarahumara *tesgüino* paraphernalia, 153

A *tesgüinada* at Rejogochi in 1981, 154

Mabel Lehi, a San Juan Paiute, using a seed beater in 1984, 166

A Kaibab Paiute summer camp in 1872, 169

A Kaibab Paiute basketmaker in 1873, 171

The baptism of a Southern Paiute man by a Mormon missionary in 1875, 178

Kaibab Paiutes meet with Major John Wesley Powell in 1873, 181

Ramón Blanco poling a reed balsa with a turtle harpoon, 1922,
     189
Seri harpoon points, 190
Ramona Casanova trimming a century plant in 1968, 194
A 1692 map drawn by Padre Adam Gilg, 199
A Seri family at Rancho Costa Rica in 1894, 201
Don Pascual Encinas with W. J. McGee in 1894, 202
Nacho Burnett and his wife, Mercedes Diaz, carving ironwood
     in 1982, 204
Seri women waiting by a tourist's car with crafts for sale, 208
Lola Blanco with necklaces for sale, 209
A Mohave traveler's kit, 221
Cocopa men and an Apache woman who was taken captive as a
     child, 223
The war leader Xantapacheria with an all-Maricopa unit during
     the Civil War, 230
A home on the Quechan reservation in about 1900, 232
A Hopi farmer in about 1900, 239
The village of Walpi on First Mesa in about 1900, 247
A Hopi bride wearing her wedding garments, 249
A young girl in a home in Sichomovi in 1902, 263

FOREWORD

All museums, but especially those in a university setting, have a threefold responsibility to the public. First, they must use the collections under their care to educate visitors in creative, attractive, and user-friendly exhibits. Second, they must ensure that the information presented is as accurate and up-to-date as possible and that the interpretations are based on scholarship of the highest standard. Third, they must make an effort to share that creativity and scholarship with those members of the public unable to enjoy the exhibit itself. In the mid-1980s, the staff of the Arizona State Museum at the University of Arizona began discussing how the museum might celebrate the hundredth anniversary of its founding in 1893. It was decided to mark the occasion by installing a new ethnographic exhibit to replace the existing thirty-year-old displays, which no longer met any of these responsibilities. We wanted to celebrate our centennial by telling the story of the American Indian peoples of the region in a totally new, historically accurate, and culturally sensitive way.

The exhibit, Paths of Life: American Indians of the Southwest and Northern Mexico, opened its first phase in November 1993 and the second in January 1995. It was the result of a productive collaboration involving Indian advisors and many members of the staff of the Arizona State Museum. It enabled the museum to meet its first two responsibilities. This book is another product of that collaboration, and we hope that it will help meet the third responsibility by sharing the creativity and scholarship of the exhibit with a much wider public.

The Paths of Life exhibit was a major undertaking, the result of almost ten years of concerted effort under the leadership of Curator of Public Programs Bruce Hilpert. He was joined by Diane Dittemore, Nancy Odegaard, Nancy Parezo, Thomas Sheridan, and Russell Varineau to form a planning team that developed exhibit themes and content.

Many museum visitors are uncomfortable with traditional ethnographic exhibits, which tend to be static. They seemed to freeze people in the past by not linking their story to the present, thereby often failing to interpret the realities of the peoples depicted. In many cases it appeared that cultures existed in spatial and temporal vacuums, never interacting with other groups and never changing through time. An unfortunate and unintended effect of this approach was that it perpetuated stereotypes of native peoples and did little to change existing views. Our challenge, therefore, was to portray the cultural vitality and historical depth of Indian cultures in the American Southwest and northwestern Mexico, the homeland of many dynamic native cultures striving to maintain their traditions, beliefs, and identities in the face of internal and external change. How could objects be combined with written words, voices, and graphics to portray the richness and complexity of this region through time without either overwhelming the visitor or being too simplistic?

With funding from an exhibit planning grant from the National Endowment for the Humanities, we began to address these problems by treating the entire exhibit as a single multicomponent attempt to explore humanistic themes, social issues, historical events, and cultural values. The planning team held marathon conceptual meetings over a two-year period to develop central themes for the exhibit based on the seminal concepts of anthropologist Edward H. Spicer, especially his ideas about enduring peoples. How did communities that had experienced so much in the postcontact period—disease and starvation, the loss of land and water, enforced labor, and even slavery—retain such a strong sense of self? We were aware that many well-informed people throughout the world still believe that Indians have either died out or have been assimilated into mainstream Mexican or American culture. How could the exhibit make it clear that people can retain their distinctive identities while interacting with the societies that surround them?

We discussed how we could use history in an ethnographic exhibit on native cultures when most of the historical documents on which we rely were written by Euro-Americans. How we could show that, while Indians were at times without political power, in important ways they were never powerless. What were the important events and processes in each society that shaped native views of themselves? What should museum visitors learn about American Indians in general and about each group in particular? We agreed that one of the main purposes of a permanent exhibit of this kind should be to confront misinformation and stereotypical attitudes.

Early discussions also centered on exhibit size and scope. The available exhibit space in the Arizona State Museum was too small to encompass all the southwestern cultures in the depth they deserve. After considerable discussion, ten tribal groups were selected: Comcáac (Seri), Rarámuri (Tarahumara), Yoemem (Yaqui), O'odham (Pima and Papago), Colorado River Yuman (Quechan, Mohave, Cocopa, and Maricopa), Nɨngwɨ (Southern Paiute), Pai (Havasupai, Hualapai, and Yavapai), Indé (Western Apache), Diné (Navajo), and Hopi.

The exhibit combines ethnographic understanding with both written and oral history, Native American as well as Anglo. All photographs and artifacts are dated as accurately as possible. Each unit of the exhibit begins with tribal accounts of origins, followed by a section on the history since European contact, and ends with a section on life today. The many different paths selected to deal with common problems and opportunities are shown to be rooted in specific philosophies and world views. The exhibit addresses directly some of the issues and problems faced by native peoples today, such as land and water rights, sovereignty issues, and the exploitation of resources by outsiders, as well as certain shared strengths, such as close ties to the land, enduring cultural traditions, vital religious beliefs, and strong family networks.

Despite these common problems and strengths, each tribal group has its own philosophy and way of life. To emphasize the distinctiveness of each group, we identified specific cultural values that have shaped and continue to provide direction for each tribe's actions. We used these as interpretive themes to highlight the uniqueness of each of the tribal groups.

However, ideas, concepts, and themes, no matter how powerful and important they may be, do not in themselves constitute an exhibit. They become an exhibit only when objects and images are joined with them to create an aesthetically pleasing and educationally effective physical presence. The Paths of Life exhibit was designed and constructed under the capable management of Curator of Exhibits Russell Varineau and Bruce Hilpert. They were assisted by many members of the museum staff, University of Arizona students, and Indian artists and interns. In reality, the entire museum was involved in the effort.

We are grateful to many organizations and individuals for the funding that made the exhibit a reality: the University of Arizona; the National Endowment for the Humanities; the Institute of Museum Services; the Arizona Humanities Council; the Getty Grant Program; the Bay, DeGrazia, Marshall, Rockefeller, Stocker, and University of Arizona foundations; several corporations, including the Arizona Bank, Dupont, the Click Automotive Group, and Target Stores; the Arizona Archaeological and Historical Society; the Watson Smith family and former Director of the Museum Emil W. Haury; and many generous individuals.

An effective exhibit is a powerful means of communication, but it has its limits. It can only communicate to those visitors who come to the museum and view it. Its message cannot reach a wider public unless museum staff members write about that message in a popular manner that has scholarly integrity. The Arizona State Museum has previously extended beyond its walls the ideas presented in exhibits on Western Apache material culture, Hopi kachinas, and southwestern women anthropologists by publishing books on these topics.

This book, which is based on the Paths of Life exhibit, is an effort to meet our third responsibility by sharing the information and interpretations of the exhibit with our larger public. Its organization is a direct reflection of the rationale for the exhibit. The book utilizes much of the material seen in the exhibit, such as artifacts, historical photographs, graphics, and special topics. The text closely follows the key themes of the exhibit, although the book, by its very nature, allows us to discuss issues and topics in greater detail than is possible in the more confining format of the exhibit. Like the exhibit, the book is an in-house product. The chapters on the different tribal groups were written by the same curatorial scholars who contributed to the development of the story line for the exhibit. Other staff members provided the illustrative material. Kathy Hubenschmidt, Helga Teiwes, and Ken Matesich supplied the photographs; Susan Luebbermann researched historical photographs; and Diane Dittemore selected objects to be illustrated and wrote many of the captions. Liane Hernandez supplied the original captions. Shelby Tisdale did research on contemporary issues, and Sue Ruiz typed many versions of the manuscript. Several anthropological colleagues served as consultants: Robert C. Euler on the Pais, Catherine Fowler on the Southern Paiutes, Charles Lamb on the Colorado River tribes, William Merrill on the Tarahumaras, and Mary Beck Moser on the Seris. At the University of Arizona Press, Amy Chapman Smith and Alan M. Schroder provided editorial supervision, while Carrie Nelson House designed the book. Carol Gifford read and edited the entire manuscript for the Arizona State Museum.

One of the most rewarding aspects of the entire Paths of Life project was the close and cordial working relationship we enjoyed with many Indian advisors, consultants, and artisans. We would like to thank them for their valuable contributions to both the exhibit and the book. They are listed here with their tribal affiliation and with any roles in addition to consultant/planning team member indicated: Kristina Ackley (Bad River Chippewa/ Oneida; intern), Cecil Antone (Akimel O'odham; Advisory Group), Clarenda Begay (Navajo; Advisory Group, midcareer intern), Alfred Blackhorse (Navajo; artist, exhibit specialist), Grace Boyne (Navajo), Jonathan Browning (White Mountain Apache), Jeanette Cassa (San Carlos Apache; Advisory Group), Betty Cornelius (Chemehuevi; Advisory Group), Gerald Dawavendewa (Hopi/Tsalagi; artist, intern, exhibit specialist), Cheryl Dayaye-Pailzote (White Mountain Apache; intern), Joseph Dishta (Zuni; Advisory Group), Paul Ethelbah (White Mountain

Apache; Advisory Group), Amelia Flores (Mohave), Fern Ho-
nawa (Zuni; midcareer intern), Louis Hood (Yavapai; Advisory
Group), Nicole Horseherder-Higley (Navajo; intern), Loretta
Jackson (Hualapai; Advisory Group), Leigh Jenkins (Hopi; Ad-
visory Group), Angelo Joaquin (Tohono O'odham), Joseph Joa-
quin (Tohono O'odham; Advisory Group), Jennie Joe (Navajo),
Weldon Johnson (O'odham; consultant for Colorado River
Tribes), Raymond Kane (White Mountain Apache; Advisory
Group), Daniel Leon (Yoeme), Merna Lewis (Salt River Pima-
Maricopa Indian Community; Advisory Group), Ramson Loma-
tewama (Hopi), Lawrence Lomatska (Hopi), Charlene Lupe
(White Mountain Apache), Felipe Molina (Yoeme), Paul Nosie
(San Carlos Apache), Edgar Perry (White Mountain Apache),
Marie Perry (White Mountain Apache), Elaine Peters (Akimel
O'odham; Advisory Group), Nancy Quade (Prescott Yavapai;
ASM Advisory Group), Gregory Redhouse (Navajo; intern),
Jocelyn Salt (Navajo; intern), Gevene Savala (Kaibab Paiute; Ad-
visory Group), Emory Sekaquaptewa (Hopi), Anna Silas (Hopi;
midcareer intern), Ronald Susanyatame (Hualapai; Advisory
Group), Clark Tenakhongva (Hopi), DeVore Thompson (White
Mountain Apache), Mary Kim Titla (San Carlos Apache), Steve
Titla (San Carlos Apache), Octaviana Trujillo (Yoeme; Advisory
Group), and Ofelia Zepeda (Tohono O'odham).

We hope that we have provided a perceptive and lasting understanding of the many paths that southwestern Indian communities have followed throughout the years.

RAYMOND H. THOMPSON, Director
Arizona State Museum
University of Arizona

INTRODUCTION

Thomas E. Sheridan and Nancy J. Parezo

American Indians have taken many different paths since they settled in the area that is now Arizona and northwestern Mexico. This book's title reflects its ties to one exploration of these paths—the Paths of Life exhibit at the Arizona State Museum—but it is much more than an exhibit catalogue. The chapters on the Navajos, Yaquis, Apaches, Pais, O'odham, Tarahumaras, Southern Paiutes, Seris, River Yumans, and Hopis parallel the storylines followed by the exhibit, but they also expand and enrich those storylines. The exhibit relies on artifacts, graphics, and space. The book remains rooted in words and images.

Many of the words spring from the sacred myths and songs of the people themselves. Each group has a unique relationship with a particular landscape. Each uses words to create and recreate the bonds that bind them to their homeland. Whether it is the Mockingbird Speech of the Tohono O'odham (Papagos) or the Deer Songs of the Yoemem (Yaquis), poetic language encodes knowledge about places and expresses a sense of place. Words are one way people inhabit the land and invest it with meaning.

Words also convey a sense of history even if that history is shrouded in myth. *Paths of Life* proceeds from the premise that all people have a history even if their history has not been recorded in conventional written records. The emphasis is therefore ethnohistorical rather than ethnological. Instead of describing the religion, social organization, and material culture of each group, we explore the ways in which American Indians have defined and redefined themselves in response to changing natural and human environments. We do not assume that there is a traditional religion, social organization, or material culture to discover. On the contrary, we examine how these ten groups have continually incorporated new symbols, ceremonies, and material items into their cultures as they have interacted with one another and the European newcomers who penetrated the Greater Southwest (the U.S. Southwest and Northern Mexico) beginning in the sixteenth century. The peoples described in this book may be enduring, but they are not timeless. Like everyone else, they have a past, a present, and a future.

We emphasize the pervasive reality of change because many otherwise well-informed people believe that Indians have either died out or been assimilated into mainstream Mexican or Anglo American culture. In Arizona and northwestern Mexico, however, there are at least as many American Indians living in the region now as there were at contact in the sixteenth century. Some reside on reservations, others in cities. Some are farmers or ranchers, others are lawyers, government officials, miners, teachers, doctors, nurses, engineers, auto mechanics, homemakers, anthropologists, or novelists. They are Republicans, Democrats, or members of the Partido Revolucionario Institucional (the PRI, or Institutional Revolutionary Party, Mexico's dominant political party). They may belong to the Church of Jesus Christ of Latter-Day Saints, one of the many Protestant denominations, the Roman Catholic Church, the Native American Church, or to organizations that spring from Uto-Aztecan, Yuman, or Athapaskan traditions of belief. Many are poor and unemployed; others are wealthy and well educated. Their diversity cannot be captured within a museum diorama or a text.

Nonetheless, we can suggest that diversity by presenting the complexities of historical change as well as the realities of modern life. *Paths of Life* seeks to portray Indians as actors, not reactors, as protagonists, not victims, as agents of change rather than foils of progress. During the late nineteenth century, when the governments of Mexico and the United States were subjugating the Indians of the Greater Southwest, most missionaries, scientists, and politicians believed that human societies

The bounty of the sea. A Comcáac (Seri) woman blows up a turtle bladder to use as a container for liquids while a Seri woman on the right cleans the carapace of a butchered sea turtle. Desemboque, Sonora, 1940. (Photograph by Gwyneth Harrington Xavier, Arizona State Museum)

could be arranged on the rungs of a ladder leading from savagery to civilization. Western culture—Christian, Victorian, and capitalist—was seen as the apogee of human achievement. Indians were many rungs down the ladder. The gulags, holocausts, and world wars of the twentieth century have demolished such evolutionary certainties, but stereotypes still persist—the Indian as Noble Savage, as Natural Man, as Victim, or as Primordial Ecologist. We hope that this book can, in some small way, provide a more balanced view of the American Indians of Arizona and northwestern Mexico.

We also want to investigate one central and fundamental theme—the persistence of ethnic identity in the face of constant change. This theme springs out of the research of the late Ed-

ward Spicer of the University of Arizona, who devoted his career
to studying how American Indians of the Greater Southwest
responded to the invasion of their territories by Spaniards, Mexicans, and Anglo Americans. Spicer's masterful synthesis, *Cycles
of Conquest: The Impact of Spain, Mexico, and the United States
on the Indians of the Southwest, 1533–1960* (University of Arizona Press, 1962), remains the definitive work on the subject
more than thirty years after it was first published. Toward the
end of his career, however, Spicer elaborated on the theme of
ethnic persistence by developing his concept of *enduring peoples*.
By *enduring*, Spicer did not mean simply surviving. On the contrary, he was talking about groups of people who have maintained a strong sense of their own ethnic identity in the face of
conquest, persecution, exile, and in some cases attempted gen-
ocide. Spicer focused most of his research on the Yaquis of Arizona and Sonora, but he compared them to other enduring peoples as well: the Basques and Catalans of Spain; the Welsh and
Irish of the British Isles; the Hopis, Cherokees, and Senecas of
the United States; the lowland Mayas of Mexico; and the Jews
across the world. He realized that he could have named a hundred more enduring peoples, including the other American Indians covered in this book.

Spicer's concept of enduring peoples is a subtle and complex
one. Even though he recognized that symbols of language,
homeland, and racial identity bind a group together, he argued
that the persistence of a people did not necessarily depend on the
maintenance of racial purity, the retention of a language, or the
continuous occupation of a homeland. The Jewish diaspora from
Palestine after A.D. 70 was mirrored by the Irish Catholic expul-
sion from Ulster in northeastern Ireland in the seventeenth century and the Yaqui diaspora of the late nineteenth and early
twentieth centuries. Most Irish no longer speak Gaelic, yet they
still consider themselves Irish rather than English. Enduring
peoples endure despite changes in the language they speak or
the place where they live. They persevere because they embrace
change and make it their own.

One of the principal ways they do so is through symbols,
rituals, and words. All of the American Indian peoples chron-
icled in this book have taken up arms against great odds to

A Yoeme (Yaqui) deer dancer, Luis Cienfuego, and three deer singers perform at Yoem Pueblo in Marana. (Photograph by David Burckhalter)

defend their homelands and their right to live and worship in their own unique ways. Yet all, at one time or another, have been conquered by superior military forces and have been scattered to reservations, refugee communities, or boarding schools far from their homes. Through those years of exile and subjugation, however, they kept alive their culture—their sense of being a people. Even though many Navajos live and work outside the Navajo Nation, the four sacred mountains continue to mark the boundaries of their spiritual homeland, the Diné Bikeyah (Navajo Country). Hopis living in Albuquerque, Tucson, or Los Angeles return to the Hopi Mesas to participate in the intricate ceremonies of their clans and religious societies. Yaqui deer singers teach the words of the *maso bwikam* (deer songs) to Yoemem children in Marana north of Tucson and in Guadalupe on the outskirts of Phoenix as well as in the Eight Sacred Pueblos along the Yaqui River in Sonora. The symbols, rituals, and words remind people of their identity and distinguish them from others. They also allow them to incorporate new ideas, rituals, and ways of making a living into their traditional social and philosophical frameworks.

In the following chapters, we discuss persistence in the face of change by focusing on some of the ways in which individual groups have maintained their identity. Rather than attempting to cover all the generally recognized anthropological cultural domains for each group, we have chosen one major theme for each chapter. The chapter on the Navajos, for example, demonstrates how the Diné have incorporated sheep—a Spanish introduction—into their culture in order to reinforce Navajo ways of relating to one another and living off the land, while the chapter on the Yaquis examines how the Yoemem have fused Uto-Aztecan and Catholic rituals and beliefs into a uniquely Yaqui religion.

But even though we concentrate on the rituals of one group and the environmental adaptations of another, we do not mean to suggest that such themes are the concern of only one people. In a book like this, we cannot be encyclopedic. Instead, we want to convey the richness of certain themes in their historical and cultural contexts while recognizing that the resonance of ritual and strong bonds with the land are vital elements of the identities of all the peoples presented here.

The themes of individual chapters were selected in consultation with respected Indian scholars who helped us to plan the Paths of Life exhibit. As Dr. Thompson's foreword makes clear, the process involved hours of intense discussion and debate, but sooner or later one fundamental question was always raised: Given all the problems your peoples have faced since the time of first contact with Europeans—Old World epidemic diseases, forced assimilation, the loss of tribal lands, and radical changes in the way you make a living—how have your peoples remained so strong and active, retaining your sense of yourselves as a people?

The answers were complex because American Indian societies are not homogeneous. There have been stubbornly conservative men and women—hostile to change of any sort and suspicious of all contact with cultures other than their own—and there have been men and women who have sought out and welcomed new ideas. The ways in which American Indians have resisted assimilation and maintained their identity are as creative and diverse as the peoples themselves. Nonetheless, some common responses emerged from our discussions with our advisors. All agreed that extended family or clan ties are strongly felt and essential to their survival. To the Native peoples of the Southwest, the American nuclear family seems like a lonely and sterile arrangement. Our advisors also emphasized a sense of community with the land, sanctioned by religious belief. Groups attempt to solve their problems by consensus; speed in decision making is less important than solidarity. Finally, the advisors believed that most American Indian societies have adapted well to change when they have been given anything like a fair chance to make the most of a new situation on their own terms.

In the following pages we explore the dialogue between persistence and change that has characterized the Indians of Arizona and northwestern Mexico during the past 500 years. We describe how the people themselves view their own origins, and we discuss how they have resisted, accommodated, and transformed the intrusions of Spaniards, Mexicans, and Anglo Americans. In doing so, we employ many different types of evidence: creation stories, sacred myths and songs, oral traditions, archaeological data, letters from missionaries, military reports, ethnographies, and the recollections of tribal elders. Each type of evidence has its own strengths and weaknesses. Creation stories and the oral traditions of clan migrations tell us much about the cultural values and self-identities of the people. In some cases, however, the symbolic rhetoric of these stories may no longer be fully understood. Moreover, indigenous traditions may or may not correspond to the speculations of archaeologists. Personal recollections of tribal members provide the most vivid cultural interpretations of events, but they often cannot extend beyond the recent past. For that reason, they must be supplemented by the observations of Spanish missionaries, Mexican government officials, and Anglo American military officers, who discuss events and personalities farther back in time. Yet the observations of outsiders are usually flawed by the observer's lack of cultural understanding or sensitivity. Consequently, they must be interpreted within the framework of more sophisticated ethnographic research, especially research informed or carried out by members of the groups themselves.

Mission San Xavier del Bac in 1863. European missionaries brought new crops, animals, and tools to the O'odham people, as well as devastating epidemic diseases. The lithograph exhibits considerable artistic license; the fortress structures in the foreground never existed. (Lithograph by J. Ross Browne)

Hualapai students at a Bureau of Indian Affairs day school in Kingman, Arizona, in about 1896. (California Historical Society, Ticor Title Insurance Collection, University of California Library)

There are many conceptions of history—history as cosmology, history as chronological narrative, history as patterns and cycles. Our approach is ethnohistorical. We have endeavored to present how the people conceptualized their own histories as well as how archaeologists, ethnographers, and documentary historians have reconstructed them. We recognize that the documentary record compiled by non-Indians is often distorted and always incomplete. As far as possible, it must be complemented by the words and interpretations of the people being written about. We also realize that history is never an objective exercise. In the words of ethnohistorian Bernard L. Fontana,

> What we call "history" is a recitation of events selected from the past, which in its most literal sense is all that has preceded the present: a rock that fell, a dog that barked, an infant who cried, a woman who coughed, a prince who was enthroned king. All historians—and on occasion each of us is a historian

A Diné woman with sheep in a pen in about 1950. (Photograph by E. D. Newcomer, Arizona Historical Society)

—select from this infinity of events those we deem worth telling. The basis of that selection provides the built-in bias of history. History, more than being a debate about the past, is an argument about the present and future. It often tells us less about what was and more about who we are. It is a tool used by all of us either to justify or condemn the status quo. It is a statement of the world either as we now perceive it to be or as we think it ought to be. The past is immutable, but history, a battleground for the public mind, is ever changing (Fontana 1994:xi).

As these chapters indicate, that battleground may have shifted from armed resistance to political mobilization, cultural revitalization, and Byzantine legal maneuvering. Nonetheless, the battles are still real battles: to reclaim lost water or lost territory, to protect sacred sites, or to create jobs and a healthy lifestyle in a world of high unemployment, drugs, alcohol, and fast food. Once marginalized on remote rural reservations or in isolated regions of refuge, the American Indians of the Southwest and northwestern Mexico are reasserting their rights to land, water, and natural resources crucial to the development of the United States and Mexico. The struggles of the Akimel O'odham to recover the water of the Gila River, of the Sonoran Yaquis to retain possession of their fertile farmland in the Yaqui Valley, and of the Hopis and Navajos to increase revenues and exercise greater control over their vast deposits of coal, oil, uranium, and natural gas are not exotic or marginal conflicts. On the contrary, the outcome of these struggles will determine the destinies of all who live in the Greater Southwest, not just of the ten groups presented here.

Southern
Paiutes

Havasupais

Colorado

Navajos

Santa Fe

River
Yumans

Hualapais

Hopis

Little

Los Angeles

Yavapais

Western
Apaches

Colorado

Chiricahua
Apaches

Maricopas

Salt

Gila

Gila

Colorado

Tucson

San Pedro

O'odham

Santa
Cruz

El Paso

Rio Grande

PACIFIC

GULF

Sonora

Seris

Pitic
(Hermosillo)

OF

Yaqui

Chihuahua
City

Guaymas

Yaquis

Mayo

Tarahumaras

OCEAN

Fuerte

Mayos

CALIFORNIA

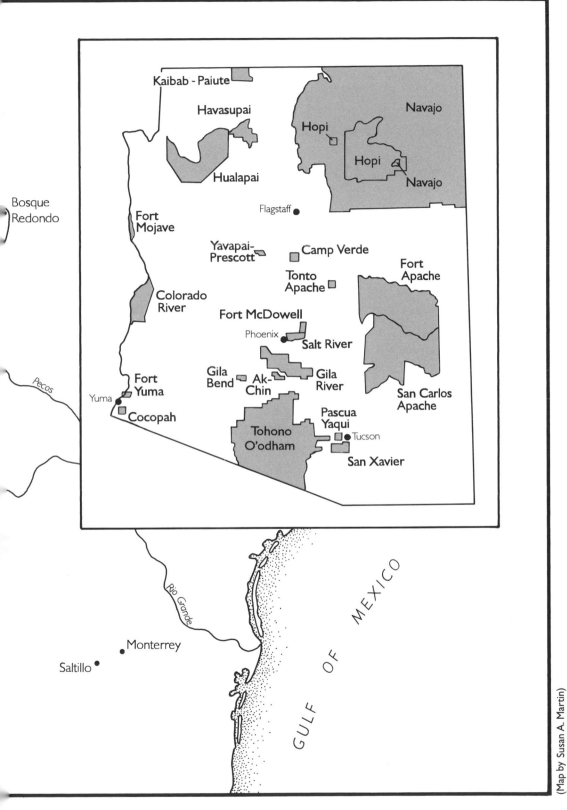

Selected tribal boundaries of American Indians of Arizona and northern Mexico in the eighteenth century. Inset: Modern Indian reservations in Arizona.

(Map by Susan A. Martin)

PATHS OF LIFE

THE DINÉ (NAVAJOS)
SHEEP IS LIFE

Nancy J. Parezo

For more than three centuries, the Diné have been pastoralists, relying on herds of sheep, horses, and cattle for their subsistence. But pastoralism is more than simply a way to put food on the table; it is a way of life that embodies important values of Navajo culture and that is a metaphor for cultural identity. It is also a state of mind that sustains those Diné who have jobs in stores and factories off the reservation and who no longer herd sheep.

Navajo history revolves around central concepts of animals, plants, the land, and journeys through a dynamic landscape. History is a story of tradition and change, of individuals and groups actively responding to a wide variety of forces and influences. The stimuli for these responses have been internal and external, and have affected all aspects of Navajo life. The Diné think of life as a journey in which walking separates activities in a world full of motion. Walking, an event important in and of itself, is a necessary element, too, in telling a story or presenting a historical account. In their mythology, Navajo orators describe walking and journeys, the landscape and places the hero or heroine passes through. Thus the Navajos walk the Beautiful Trail, and their history is a story of individual and group journeys—both through evolving mythical worlds and across the present landscape—in search of harmony, order, and peace.

Diné history begins with the creation of the world at The Place Where the Waters Crossed. The Navajos tell of four worlds through which their ancestors journeyed before emerging into this, the Fifth World. Each of these worlds was colored and in-

A shepherd and his flock at Pine Springs, near Chinle, in 1970. (Helga Teiwes, Arizona State Museum)

habited by primordial peoples who were incorporated into Diné society. People tried to live in peace, but disharmony and disorder forced them to ascend to higher worlds in their search for an orderly universe. Ultimately, they moved to the Fifth World, emerging in the center of Navajo sacred land only to discover that the world was covered with water and inhabited by monsters. The Holy People came to terms with Water Monster, from whom Coyote had playfully stolen a child. When the child was returned, the waters receded, and the world of the Diné was ready to be formed.

The next stage of Diné history involved preparing the land and making it beautiful by eliminating chaos. The stars, moon and sun, food animals, plants, and mountains were created, day and night were distinguished, and time was divided into four seasons. Changing Woman was born of the mingling of darkness and dawn, and her children—Child Born of Water and Monster Slayer—defeated the monsters and made the world safe. Only when all was ready did First Man and First Woman build a

CREATION STORY
THE GATHERING OF THE CLANS

Nancy J. Parezo

When Changing Woman left for the west, her younger sister, White Shell Woman, returned to her home in the San Juan Mountains. But after five days White Shell Woman longed for company. Talking God saw her plight and brought all the Holy People to her house. Changing Woman brought two sacred blankets of dark embroidery and white embroidery. Talking God brought two buckskins, and First Man and First Woman brought two perfect ears of female and male corn.

With ceremony and prayer, Talking God pointed the corn in the four cardinal directions. He placed them on the blanket under the buckskins. The Wind breathed on the corn four times, and the white corn changed into a man and the yellow into a woman. Rock Crystal Boy gave them minds, and Yellow Corn Girl gave them voices. When the Holy People were finished forming these Earth Surface People, they went into White Shell Woman's home, and the Holy People departed for their homes.

In four days, Talking God brought Ground Mist Girl and Mirage Boy to marry the first siblings. Each couple had one boy and one girl. From these people are descended the Tséníjíkiní clan, the Honeycombed Rock People. Because Talking God pointed the corn in the four directions, the Navajo people never live like the Pueblo people. They migrate constantly from place to place and are scattered across the land.

One day, the Honeycombed Rock People saw a fire far off in the distance. They searched for the source for many days, and with the help of the Wind they found a small encampment of Earth Surface People. The people rejoiced and asked, "Where do you come from?"

"We come from Tséłigaii íí'áhí, the Place Where the White Rock Stands," the newcomers answered. "We have survived only on ducks and snakes, prairie dogs, and whatever berries and seeds our women and children have been able to gather."

The people of the Honeycombed Rock clan asked them to live with them because they had food. "You are welcome to travel the path we travel, to rest where we rest, to hunt where we hunt, and to share that which we have." So they brought the people to White Shell Woman, who named them Tsétł'áhnii, or Rock Corner Place People, after the place where they had met. She called all of them her children.

Soon other clans came and were named. Then Talking God took White Shell Woman to live in Tsegi Canyon at the Place of the Mountain Sheep. The people did not want to see her go, but she told them not to be sad: "I will come often to see you and I shall watch over you. Your need for me is no longer great. You are growing stronger as a people and learning to control the things of this world. Look for me when you see the gentle showers of female rain."

Now when new people came, they named themselves. Some came from other groups-— Ute, Zuni, Pueblo, Mexican, and Apache. Some were formed by Changing Woman when she was lonely. Others had no known history, having survived the monsters. All had long journeys and faced many hardships and adventures before finding the right place to live. Sometimes they dwelled apart for a while, but eventually they joined the Diné and behaved as if they were one. Each brought new skills, which they shared. From that time on, the Diné flourished as they hunted and grew food, living simply here on the surface of the Fifth World.

(Adapted from Zolbrod 1984)

hogan. In the hogan the Holy People (immortal beings who travel by following the path of the rainbow and sunray) created Earth People, sent them on migrations, and taught them how to be human—how to bathe, to cook, to understand, to contact the Holy People in the proper manner, to behave properly, and to make clothing, pottery, baskets, and tools. The Holy People stressed the importance of ceremonies and prayer to insure that the world and the Diné's travels would be harmonious and orderly, for as Irene Stewart has said, "When our Blessingway is forgotten, our elders say the earth will be destroyed by fire" (1980:52). Earth People thus have grave responsibilities. The Diné were taught to search for *hózhǫ́*, a state in which all is beautiful, harmonious, and good, and how to put the world in order when individuals or evil things place it in flux. All life is interdependent, and Navajos respect all beings, not just the Holy People, for their special characteristics.

Navajo culture is marked by flexibility and evolution as well as continuity, which is fortunate, since the Beautiful Trail has not always been easy but has instead been filled with many trials and adventures. For centuries, Diné have adopted many things that complemented and enriched their lives and have rejected those which did not. They have, in turn, offered new ideas to the peoples with whom they interacted. For example, the Navajos learned to farm from the Puebloan peoples. So prolific did they become that Franciscan missionary Zárate Salmerón referred to them in 1626 as the Apaches de Nabajú, "the Apaches who practice agriculture" (Correll 1979:30). From the Spaniards, the Diné obtained sheep, goats, and horses. Tending flocks became the mainstay of their life in this arid land. Navajo cultural flexibility, with its underlying core of traditionalism, has allowed the Navajos to prosper—indeed, to all but reinvent themselves when necessary. They have grown by accretion, incorporating new peoples and new ideas and modeling them into that which is distinctly Navajo.

The People's Journey to Their Homeland

The Navajos have not always lived in their current location. The Holy People guided the Earth People on long treks to their present home. Some 800 to 1,000 years ago, family groups of Diné and other Athapaskan groups (like the Apaches) began to move slowly southward from their home in the cold, dark climates of the north and west to Dinétah, a land of warmth and sunshine in the Southwest. This voyage is mirrored in the Navajos' emergence tradition, which tells of their travels from the cold, dark worlds below to this one of light and variegated color. To anthropologists, language and customs confirm the Navajos' affinity to the many Athapaskan speakers in both Canada and along the northwest coast of the United States, as well as to the various Apache groups of the Southwest. Like all hunting-and-gathering peoples, the Diné lived in relatively small, flexible bands and extended family groupings, meeting occasionally to discuss issues of concern. Religious leaders concentrated on curing the sick and insuring abundant food. The Navajos made conical skin houses, baskets, harpoons, sinew-backed bows, snowshoes, and skin clothing decorated with porcupine quills, and they used dogs to carry loads and pull travois.

This period of Navajo history consists of clan migrations. All Navajos belong to a matrilineal clan composed of large groups of relatives identified with a common female ancestor. Navajo clans include Tódich'íí'nii (Bitter Water People), Kinyaa'áanii (Towering House People), Ashiihi (Salt People), and To'aheed'ííni (Water Flowing People). Other clans came into existence later as the Diné grew and as people from other societies married into the tribe. A Navajo is Born To the mother's clan and Born For the father's clan. Even though the Navajos consider themselves to be a people of many clans, they feel an underlying unity because they share the same beliefs. They are, as many Navajos will say, a people who walk with beauty all around them.

The Navajos' clan migrations on the Beautiful Trail brought them to the Southwest. The Diné settled in sparsely populated areas near Pueblo peoples in the canyons of what is now southern Colorado and north central New Mexico. Although the clans at first occupied specific areas and became associated with geographic features, members soon moved throughout the Diné homeland. No one clan was superior to another. The Diné did not constitute a single political entity, a cultural fact that Spaniards and Anglo Americans failed to comprehend. Misunderstandings

resulted that had important ramifications for the Navajos. While
the Navajos shared customs and a common language, their emergence as a tribe and a nation came only in recent times.

The Navajos quickly became part of Diné Bikeyah (Navajo
Country) inside the four sacred mountains—in the east, White
Shell Mountain (Blanca Peak in central New Mexico); in the
south, Blue Bead Mountain (Mount Taylor in northwestern
New Mexico); in the west, Abalone Shell Mountain (San Francisco Peak in north central Arizona); and in the north, Obsidian
Mountain (Hesperus or La Plata Peak in southwestern Colorado). As each clan realized that it had found its homeland, the
Navajos imbued the features of the land with significance. They
instilled in their children a sense of belonging sanctified by tradition.

Learning to Live in the Southwest

Changes in Diné culture occurred as Navajos learned to live in
the arid canyon lands, developing a highly successful adaptation
to their environment based on a diversified economy. The Diné
admired the reliable sources of the food the Pueblo peoples grew
and saw that farming could make their lives easier. Agriculture
emerged as a cornerstone of the economy, supplemented by
hunting and gathering. Corn became very important in Navajo
life for it was more than food. Corn pollen was used in ceremonies to symbolize fertility and prosperity, and the growth of
individuals and the Navajo people came to be likened to the
growth of a corn stalk.

The coming of the Spaniards offered new opportunities as well
as hardships for the Diné. Navajos acquired sheep and horses,
thus adding pastoralism to their economic base and enabling
them to become a dominant force in the Southwest. The Navajos
soon came to think of themselves as pastoralists. Since the 1700s,
sheep and horses have provided food, wool, and transportation.
Animals are given to others as signs of generosity and are used as
payment for services, such as a singer's fee to perform a curing
ceremony.

A pastoral life requires seasonal movements to grazing lands
in the mountains in summer and the lowlands in winter. Thus

the Navajos did not build permanent adobe houses but developed hogans, single-room circular or hexagonal structures of wood covered with mud. The entrance always faces east to pay respect to the rising sun, and Navajos locate them near relatives, who work together, pooling their labor. Until very recent times, local groups—ten to forty families under the direction of a Hózhǫ́ǫ́í Naat'ááh (headman)—were economically and politically self-sufficient. A man of prestige who knows Blessingway, the headman led his people by persuasion, ability, wisdom, and eloquence, characteristics that the Navajos value in their tribal leaders. Their pastoral life also meant that the Navajo people were scattered over a large territory, minimizing death from contagious disease. The smallpox epidemic of 1780, for example, had little effect on the Navajos, unlike the Hopis. But this did not mean that disease has been unimportant in Navajo history. The influenza epidemic of 1918–19 was devastating—more than a fifth of all the Navajos died.

Navajos generally tried to avoid the Spaniards, with whom they had a precarious relationship. The seventeenth and eighteenth centuries saw alternating periods of peace and hostility because the Diné began raiding and because the Spaniards attempted to convert them to Catholicism. The Spaniards also conducted a slave trade that resulted in the theft of many Navajo children, a major cause of unrest. Navajos interacted more regularly with the Pueblo peoples, although even with them relations alternated between friendship and war. Pueblo groups quickly noted that the Navajos were skillful warriors who should be treated with respect. In 1630, Franciscan Fray Alonso de Benavides described an elaborate ritualized encounter with a Tewa group who had traveled to Navajo homes, which demonstrated a sharing of religious concepts and long-established trading relationships between particular families. Navajos also visited Pueblo settlements during dances and traded meat, hides, salt, alum, and later blankets in return for clothing and pottery. The Diné's reputation as great traders grew, as did recognition of their skill in working buckskin, making baskets, and weaving cloth. By the mid-1700s, Navajo blankets had come to be prized throughout the Southwest, on the Plains, and in Chihuahua. In the estimation of New Mexican Governor Fernando Chacón, writing in 1795, Navajos worked their wool with more delicacy and taste than did the Spaniards. The manufacture and trading of goods was a crucial part of the Navajo economy. By 1800 the Pueblos were relying heavily on Navajo hides, blankets, and clothing, and the trade at Pueblo fairs was immense (see pl. 1).

The Navajos' interactions with the Pueblos periodically intensified. They were allies during the Pueblo Revolt of 1680, and they shared in the captives taken. Following the reconquest in 1692, many Pueblo refugees lived with the Navajos in Dinétah. Early-eighteenth-century sites in the area of present-day San Juan, New Mexico, for example, consist of clusters of hogans and small masonry pueblitos that were defended against Ute, Comanche, and Spanish attacks. Pueblo refugees (Tewa, Keresan, Jemez, and Zuni) became incorporated as named clans and brought with them the knowledge of many skills that the Navajos found useful, including a rich ceremonial lore, irrigation agriculture, and pottery making. But not all their ideas were adopted. The Navajos never used a kiva; ceremonies were conducted in the hogan.

Increasing Ute raids on their eastern settlements led the Navajos to abandon the Upper Chama River area and move west into unoccupied lands and Havasupai and Paiute territory, absorbing some Southern Paiute groups along the way. Navajos also retaliated by attacking Spanish homesteads and raiding for livestock, and they are credited with being a major factor in the abandonment of San Gabriel, the capital of New Mexico. Raiding and counterraiding followed a cycle interlaced with agriculture, herding, and trading. The Spanish noted that the Navajos were especially given to raiding during the winter months, not during periods of planting, growing, and harvesting. They conducted raiding and warfare under the leadership of their Hashkééjí Naat'ááh (war chiefs), individuals noted for their valor and wisdom. Neither side won the undeclared war, and after Mexico became independent of Spain, New Mexicans formed a private army to fight the Navajos. In the continuing warfare, thousands of Navajos and other Indians were captured and sold as slaves in New Mexico. Individual Navajo groups debated what to do about the situation, which resulted in a split in the tribe. By 1820 one group, the Diné 'ana'í, who favored peace, had detached themselves and had settled at Cañoncito.

Navajo farms and grazing lands were threatened anew by the arrival of the "New Men," Anglo American settlers and soldiers, in the 1850s. Trade with Anglo Americans gave New Mexicans more firearms, and the earlier pattern of intermittent peace and war intensified. Treaties were proposed and concluded, but they were quickly broken on both sides—by New Mexicans who appropriated Navajo grazing lands and captured slaves, and by Navajos who continued to raid. New Mexicans labored under the delusion that headmen and war leaders were chiefs with coercive powers over all the Diné, that is, individuals with whom binding treaties could be made and enforced. Most Navajos, however, had no knowledge of these agreements.

Antagonism and misunderstandings were rampant, and the situation worsened in the 1850s and 1860s. With the murder of Narbona in 1849 and the Fort Fauntleroy Massacre in 1861 (in which Navajo women and children were shot while obtaining rations), the Navajos, but not the New Mexicans, were required to make restitution for damages from raiding expeditions, and the Navajo, but not the New Mexican, slave raids were curtailed. The Navajos justifiably mistrusted the u.s. Army and ignored its demands. Skirmishes increased. The army could not defeat the Navajos, and settlers clamored for access to Navajo lands as well as revenge. With the Civil War on, the Union wanted peace so that the region could be opened for mineral exploration and settlement. It also wanted to free the increasingly bored troops to fight the Confederacy. Slave traders, however, wanted the Navajo war to continue because it increased opportunities for raids. Even after the Civil War, Navajos were held as slaves; the superintendent for Indian affairs in New Mexico himself owned six.

The Long Walk: Years in Exile

In 1862 General James Carleton decided to launch a major campaign to relocate the Navajos to a "spacious tribal reformatory." Carleton's objective was to teach the Navajos "new habits, new

ideas, new modes of life" by transforming them into sedentary
farmers living in twelve villages located a mile from each other
and under the direction of a military official. Navajos would be
compelled, Carleton declared, "to teach their children how to
read and write, teach them the arts of peace, teach them the
truths of Christianity. . . . The old Indian will die off. Besides,
you can feed them cheaper than you can fight them" (L. Kelly
1970:21).

Carleton first had to defeat the Navajos, however, and to do
this he undertook a scorched-earth campaign. He told Colonel
Kit Carson that if the Navajos refused to surrender, "the men
are to be slain whenever and wherever they can be found. The
women and children may be taken prisoners" (L. Kelly 1970:20).
Although this ultimatum was never conveyed to the Diné, Carson led Ute, Pueblo, and New Mexican volunteers on a campaign
of destruction. They burned cornfields, felled peach trees, poi-
soned wells, and shot livestock during the harsh winter of 1863–
64. Several thousand Navajos scattered to isolated canyons on
Black Mesa, near Navajo Mountain, and along the rim of the
Grand Canyon. Many other Navajos, faced with starvation, surrendered. Another 300 were killed, 87 were wounded, and 703
were captured.

The Long Walk (the forced migration and the detention itself,
which lasted from 1864 to 1868) is one of the most significant
events in Navajo history. More than 8,300 men, women, and
children marched 250 miles to desolate Fort Sumner in eastern
New Mexico, where recently defeated Mescalero Apaches were
being detained. Tales abound of the suffering: "The trip was
made on foot. People were shot down on the spot if they com-
plained about being tired or sick, or if they stopped to help someone. . . . There was absolutely no mercy" (Curly Tso, quoted in
Roessel 1984:103). Stragglers were captured by New Mexicans
and sold into slavery. More than 300 individuals from the first
group of 2,500 perished on the journey.

Fort Sumner (also called Bosque Redondo) was a concentration camp where more than 9,000 people lived on land that could
not support them, separated from their homeland, relatives, and
livestock. Disease, inadequate housing, malnutrition, mineral-
ized water, poor soil, a lack of firewood, cutworms, wolves, and

Soldiers count Navajo prisoners at Fort Sumner in 1864 or 1866. The Long
Walk of the Diné to Bosque Redondo was one of the most traumatic episodes
in Navajo–U.S. relations. (U.S. Signal Corps Collection, National Archives, courtesy of
the Museum of New Mexico).

failed crops plagued the Navajos. The government planned
poorly for the "experiment." Some food rations never arrived,
and those that did were often spoiled. People could neither hunt
nor gather, and Comanches stole what little they had. Several
thousand people starved or died of dysentery, and the high death
rate further split families. After three years, Navajos began to
escape; others continued to cry to be released. As Irene Stewart
recounts Greyeyes' experiences, "They begged the colonel in
charge to let them return to their own country. . . . One old
woman said, 'Please, my son, let us go. I am lonesome for my
land, my dear Chinle and Canyon de Chelley' " (1980:55).

The experience of the Long Walk had a major impact on Navajo attitudes toward the federal government. It was a watershed between the past and the present. Generations later, Diné
still speak with pain of the experience. Forced assimilation did
not work: Navajo culture was not destroyed, but it did change.
The Navajos' exposure to new clothing, tools, and weaving materials had a lasting impact, and a symbol of Navajo identity
developed. A V-necked shirt of calico, white muslin trousers,
and a Pendleton blanket became standard men's clothing, and by

the 1880s most Navajo women had adopted the long skirts and blouses worn by Anglo women at Fort Sumner (see pl. 2). This costume developed into the full pleated calico skirts and colorful velveteen blouses associated with Navajo women, an ensemble in which they take great pride today. Even out of adversity, the Navajos developed something beautiful.

After the Long Walk: The Return Home

In 1866 Carleton was relieved of command, and a government study concluded that the relocation experiment had been not only a dismal failure but an unjust, cruel, expensive, and poorly managed operation as well. Since federal officials noted that few Anglos wanted to settle on Navajo lands under the Homestead Act of 1862, negotiations began. Barboncito, a respected headman and a persuasive speaker, argued for Navajo interests in the peace talks, which led to the treaty that restored to the Navajos a small portion of their lands, some 3.5 million acres. As Barboncito told General William T. Sherman on May 28, 1868,

> The bringing of us here has caused a great decrease in our numbers. Many of us have died, also a great number of our animals. . . . When the Navajos were first created, four mountains and four rivers were pointed out to us, inside of which we should live, that was to be our country. It was given to us by the first woman of the Navajo's tribe. . . . I hope to God you will not ask me to go to any other country than my own. We do not want to go to the right or left, but straight back to our own country. (Correll 1979:130–31)

The Navajos agreed to live peacefully and not to oppose a planned railroad, and the government agreed to provide them with agricultural supplies, formal education (a teacher for every thirty students), 1,000,000 pounds of maize, 15,000 sheep, and 500 head of cattle. The sheep became the key to Navajo recovery.

Only 4,000 Navajos had survived to walk home, but Barboncito predicted that "after we get back to our country it will brighten up again and the Diné will be as happy as the land. Corn will grow in abundance and everything will look happy" (Correll

The treaty of 1868 between the United States and the Navajos, which allowed the Diné to return to parts of their homeland. (Navajo Nation Museum)

1979:131). Re-establishment was not easy even though the Diné were again within the four sacred mountains and were made safe through prayer and ceremony. Although their sheep were delivered in 1869, agricultural tools and rations were slow in coming, and many people faced starvation during the first reservation winters. But people worked hard, and by 1878 the original 15,000 sheep had multiplied to over 700,000. In 1880, more than 1.1 million pounds of wool were marketed, and another 100,000 pounds were retained for home use. With this type of expansion, the Navajos could not remain within a reservation only one-tenth the size of their former territory. Throughout the late nineteenth century, conflicts once again erupted between Navajos and non-Indians on the eastern part of Diné Bikeyah, and the Navajos moved westward. As a result, the reservation was

Sheep being dipped after shearing to kill ticks and other parasites at Round Rock. Family groups usually worked together on big jobs like shearing and dipping. (Photograph by J. L. Rush, Museum of New Mexico)

Navajo men prepare wool for market in about 1935, during the period when livestock reduction became the second major trauma in Navajo–U.S. relations. (Photograph by A. V. Kidder, Arizona State Museum)

expanded several times between 1878 and 1933. Even so, a new problem was emerging: the land was rapidly becoming over-grazed.

In 1891 a government survey counted 19,000 horses, 500 mules, 9,000 cattle, and 1.5 million sheep. The Navajos felt prosperous, for sheep were wealth. "Our sheep are our children, our life and our food," said the wife of John Chief (Roberts 1951: 108). Mutton was an important dietary staple, and the Diné had more protein in their diet than any other southwestern group. Added to this base were Arbuckles coffee, potatoes, bacon, canned tomatoes and peaches, and flour—all obtained from trading posts. Sheep meant the assurance of never starving, a hedge against poverty that was important physically and psychologically to people whose environment could easily turn harsh.

The Navajo economy centered on livestock, rations, subsidies, trade, and wages. Life matched the needs of the flocks, which required a highly defined yearly schedule of movement. In the spring, sheep were shorn and the wool was taken to one of the

A Navajo silversmith in about 1910. The Diné learned silversmithing from Mexican smiths in New Mexico in the nineteenth century and then developed their own distinctive styles. (Photograph by Ben Wittick, Museum of New Mexico)

trading posts scattered throughout the reservation. Wool was traded in bulk and exchanged for articles from the outside world: food, needles, thread, enamelware, knives, shears, saddles, bridles, calico and velveteen, hats, and shoes. Traders shipped the wool to the East, where it was used for the manufacture of carpets, hats, coats, padding, and army uniforms. Navajo weavers also produced rugs that, like silver jewelry (see pl. 3), were bartered or used as collateral for loans during hard times. Traders such as Lorenzo Hubbell, C. N. Cotton, Thomas Keam and John Moore encouraged the Navajos to produce and market their woven goods just as they encouraged them to increase their herds (see pl. 4).

Sheep provided psychological as well as material security. They were mobile wealth. Sheep and horses were "soft goods." Better than money in the bank, animals paid interest each spring with lambs and foals—they replenished themselves. As Changing Woman told the Navajos after she created sheep and goats, "By means of these you will be able to live. . . . It represents your pet from the tip of which fabrics and jewels of every description

The Hubbell Trading Post, Ganado, Arizona. Lorenzo Hubbell, who purchased the post in 1878, encouraged the weaving of high-quality Navajo rugs and blankets. The post was designated a National Historic Site in 1967. (Photograph by Helga Teiwes, Arizona State Museum)

NAVAJO WEAVING

Nancy J. Parezo

For more than a century, Navajo pastoralism has been a focal point for communication, both within the Navajo community and between the Navajos and the outside world. Perhaps more than any other single item, Navajo weaving has stimulated contact, knowledge, and understanding between the Diné and other American Indians, Mexicans, and Anglo Americans, for trade has always enabled people to communicate. Navajo weaving conveys Diné concepts of beauty and craftsmanship (*nizhoni*).

Much has been written about how weavers adapted some of their blanket designs in the past to satisfy the tastes of their nonreservation customers, but weavers always reinterpreted and enhanced the beauty of the customers' concepts. Today the popularity of Navajo weaving has grown with the expansion of the ethnic art market. Contacts between the Navajos and the Anglo American world have increased as a result, and Anglo Americans recognize the inherent beauty of the designs in the weavings. But there are other messages. For example, when weavers demonstrate their art at the historic Hubbell Trading Post at Ganado, Arizona, they convey the dedication, creativity, and sense of beauty that are the essence of Navajo weaving. Beauty is in the making, and demonstrating the process of weaving is in many ways more important than the finished product.

For the Navajos, beauty is integrated with good, and moral good tends to be equated with aesthetic good: that which promotes human survival and happiness is seen and experienced as beautiful. Beauty is in the nature of things as well as in people; it is the natural state of affairs. The land is beautiful by definition because the Holy People designed it to be a harmonious, happy, and healthy place. For beauty to be maintained, it must be expressed in actions such as the creation of art. Thus, making a rug is in itself an act of beauty. "The piece is merely the vehicle whereby beauty, *hózhǫ́,* is transmitted from an artist, who is himself or herself in a state of beauty, to a recipient or audience, who will in some measure be brought into a state of beauty through viewing, wearing, or appreciating what the artist has been doing and made" (Lincoln 1982:40–41). The Navajo concept of beauty is an extremely active one, and people weave because creativity brings enjoyment and wholeness, as well as economic rewards.

The sense of beauty is conveyed to Navajos through the story of Spider Woman, a mythical teacher and a source of inspiration. Many Navajo stories contain descriptions of weaving that go back to the first worlds. Long ago, Spider Woman, who lives high atop Spider Rock in Canyon de Chelly, taught Navajo women how to spin thread and weave blankets on a loom built by her husband, Spider Man. She instructed the women so that the Diné would have clothes and not be cold. In another sense, Spider Woman is a symbol of the tradition of passing on knowledge from mother to daughter. A young girl sits at the feet of her mother or aunt beside the loom, watching and learning. Her mother names the parts of the loom, just as Spider Woman named them for the Navajo people. The young girl learns the prayers to Spider Woman that will help to ensure a beautiful creation. Through this mechanism, the Diné convey their concept of themselves as a people to outsiders.

Navajo women weaving at their camp in Keams Canyon, Arizona, in 1893. Charley weaves a blanket on a horizontal loom; a child, Nedespa, cards wool; and two unidentified women spin wool and weave on a belt frame. (Photograph by James Mooney, Smithsonian Institution, National Anthropological Archives)

will begin to sprout, thus making life possible for you. And you will plead with them, pray to their feet, their head, pray to their bones" (Wyman 1970:246). A family's public image, prestige, and status were linked to the size and well-being of the herd. No one was respected in the community or elected to tribal office if the family's livestock were not cared for properly.

For the Diné, goodness means productivity, dependability, helpfulness, and cooperation. Livestock made these values tangible. Although sheep were owned by individuals, extended families combined their herds and shared grazing territory. As Dr. Jennie Joe has remarked, "Sheep keep our families together" (1988). Caring for sheep and livestock was the concern of everyone in the residential unit—everyone had a job to do. Sheep also validated marriages and made kinship affiliations tangible. It was not unusual for individuals to have sheep in more than one herd. The flocks were a focus for social integration and a means of teaching the values of Navajo culture to the younger generation. Caring for livestock, sharing jobs within the family, and learning traditional lore were all part of becoming and being a Navajo. Those without sheep felt useless.

Mastering the skill of weaving symbolized how the gift of wool was transformed into beautiful forms for the benefit of the family. As Irene Stewart told anthropologist Mary Shepardson:

> I can remember how the Navajo women kept their people alive by weaving rugs and caring for sheep and goats. In this way my stepmother kept us from starving. She wove all the known types of rugs, trading them for food and clothing. She wove for other Navajos who admired her weaving. I can remember how she got 25 goats and sheep by weaving a fine saddle blanket for a rich man with much stock. (Stewart 1980:42)

While weaving has been described as a woman's job, it was more than that. According to Joe, "I don't think weaving is considered only a female activity—only at a very narrow point, the point of thinking of a pattern and actually weaving. But any other time both before and after it is a family activity. . . . My grandfather rolled yarn in the evening. Sometimes the men went and got the dye so the ladies could dye. Kids help also. Weaving brings families together" (1988) (see pl. 5).

Weaving also cemented family ties because producing a blanket to keep someone warm was a woman's gift to her husband or son. "A saddle blanket is a useful thing. You can give it to your husband or son. It helps to solidify the kinship ties" (Joe 1988). Women also weave traditional dresses as presents for their female relatives or give rugs to the people they care for so that they can live surrounded by beauty and creativity. Navajo arts are inextricably woven into the social fabric of everyday life.

The Navajo Sheep Reduction

Navajos were not isolated. Government schools, missionaries, wage work on the railroad and in towns along the reservation's borders increased contacts with the outside world. Many Navajos worked as shepherds or cowboys for Hispanic and Anglo American ranchers and as builders of houses and roads. A pattern developed of men and women going off the reservation for a time, working for several months or several years, and then returning to family lands and herds. But land remained a problem. By 1930, 45 percent of the topsoil had washed away and animals were scrawny. Government officials estimated that the land could carry no more than 600,000 sheep and called for conservation reforms. By 1933 more than 1.1 million sheep and goats were being grazed on the Navajo Reservation, overwhelming the capacity of the land. That year, the Bureau of Indian Affairs (BIA) began a program of stock reduction as a solution aimed at reducing stock by 45 percent by the 1950s. The program was to be voluntary and was to consist initially of a 10 percent across-the-board reduction, which large herd owners could meet by culling weak sheep but which would force small owners to sacrifice good stock. Livestock would be purchased with Federal Emergency Relief Administration funds to feed people living in other parts of the Depression-ravaged country. To convince the Navajos of the necessity and value of the program, John Collier, the commissioner of the BIA, promised financial compensation and told the Navajos that, in return for their

LEARNING TO BE NAVAJO

Nancy J. Parezo

Brian Francis teases an old ram at his mother's family's home in Blue Gap, Arizona, in 1974. As Irene Stewart said, "Everything we learn is from caring for the sheep" (1980:23). (Photograph by Helga Teiwes, Arizona State Museum)

Diné children are valued members of their families and are immediately incorporated into the group. At birth, babies become the center of attention. As Max Hanley remembered, "After I was born I was wrapped in the sheepskin and placed on the west side of the fire in the center of the hogan" (quoted in Johnson 1977:18). Then the transmission of custom begins through prayer, example, and celebration. The child's first laugh is celebrated by the family by hosting a feast of goat and mutton to insure that the child will not be selfish. Acts of gift giving by relatives "endow the child with good qualities of character, especially generosity and hospitality. After this the father makes a cradleboard from a perfect tree" (Stewart 1980:12). The parents carry the baby on the cradleboard, and when they set it down they place it so that the baby can easily watch their actions, for it is by watching and listening that children learn to be Diné.

Individuals are raised in an extended family setting, and it is not uncommon for a child to spend part of his or her childhood living with a grandmother, aunt, or uncle. All these individuals teach by example. Boys learn hunting, tracking, and ceremonial skills; girls learn cooking, weaving, and domestic skills. All learn to herd sheep, a very enjoyable and important experience that makes them productive members of the household at a very early age. The sheep incorporate children into Navajo life, the communal economy, and the residence group.

Parents give their children lambs when they are the "right age," when they are old enough to be responsible. This is an important point in a youngster's life because it marks the first step on the path toward knowledge and wisdom, and it is a sign of maturity and trust. It also marks the beginning of a person's economic future, for it is the beginning of an individual's herd. Herding sheep teaches children to pay attention to the world around them, to accept responsibility, and to learn the skills that are important for traveling on the Beautiful Trail. As one man described his activities,

> Early in the morning we take the sheep out of the corral. I sing a song and open the gate. When the sheep are half out my song is half finished. When they are all out I stop my song. . . . When you are out herding there are songs for the protection of the sheep and to make them increase. (Quoted in Kluckhohn and Leighton 1946:70)

As Irene Stewart recalled, balance in activities and industriousness is stressed:

A Navajo boy or girl is told not to play too much for too much playing will cause one to become a negligent and neglectful person. So I learned to herd sheep, prepare wool, and weave. . . . I enjoyed helping with the poultry and herding the sheep. . . . Everything we learn is from caring for the sheep. (Stewart 1980:23)

Sheep herding instills important Navajo values: rugged individualism and love for personal freedom, generosity and helping relatives and friends, steadfastness, and self-sufficiency. Caring for sheep is a way of teaching the younger generation responsibility and what it means to be Navajo. It teaches them that all creatures must be nurtured and that it is the responsibility of all Navajos to treat the land and all beings with respect.

voluntary cooperation, they would receive more land on the eastern end of the reservation, soil conservation programs, water development programs, day schools, and jobs on Emergency Conservation Works programs. The Navajos, who had argued for years that they needed more land, considered this a minimally acceptable binding agreement by Collier, but Collier could not control Congress, which had to approve all BIA programs. New Mexico politicians kept what has been called the checkerboard area from becoming part of the reservation. Navajo flocks were destroyed by the federal government, with little or nothing given in return. Navajos interpreted this as paternalism and as evidence that the government wanted to destroy them.

The Navajos strongly expressed their feelings about their flocks: "For the white people, the good old dollar is where they get their substance of life, and the Navajo get their substance of life from the goats and the sheep" said Fred Nelson in 1934 (Parman 1976:55). "Why should the Government rob us out of our sheep? It is our money," said Mrs. Y. N. Yazzie of Toadlena in 1937. "Although we were told that it was to restore the land, the fact remains that hunger and poverty stood with their mouths open to devour us," noted Dan Phillips. "Sheep is life. Who can live if their life is taken away," said a woman from Smoke Signal, Arizona (Wood, Vannete and Andrews 1982:1). To the Navajos, reduction in the size of flocks was a sign of laziness and imminent starvation.

Resistance grew. By 1937, government agents had become crusaders for reduction. Animals were forcibly taken away or killed, often without explanation, according to the Navajos. Government officials disagreed. Collier declared that he had consulted with the tribal council already and would act to enforce grazing regulations and erosion controls even as he withdrew promises of land restoration and expansion. Government officials refused to consider Navajo values in their decisions or to work closely with local groups; some even threatened to jail people who disagreed with them. To a society whose self-esteem was so closely tied to animals, the decision to slaughter sheep and goats for no apparent reason was cruel and sacrilegious. Some families were made destitute by the process, because their

Laura Ann Nez weaves a rug at her sister's hogan at Blue Gap, Arizona, in 1974.
(Photograph by Helga Teiwes, Arizona State Museum)

herds had been so small to begin with that they could not survive
on the remainder. As Irene Stewart remembered, "Many Navajos sold below their quotas and others who had few sheep
ended up with even fewer. My father ended up with none at all"
(1980:46). Civilian Conservation Corps (CCC) programs could
not take their place, and welfare was initiated to keep some
Navajo families from starving. Chee Dodge, leader of the "traditionalists" and chairman of the first Navajo Tribal Council in
1923, went to Congress to obtain more land, limit the reduction,
and stop the onset of a welfare state, but he was only partly
successful.

The emotional impact of the slaughter left deep psychological
scars and increased distrust of the government, which have had a
far-reaching impact. When the Indian Reorganization Act (IRA)

of 1934 was passed, each tribe was given the opportunity to write a constitution and reorganize its government in order to increase self-determination (though it was still under the watchful eye of the BIA) and allow it to function as a corporation. Passage of the IRA would have allowed the Navajos to organize a council and courts, and to borrow money as a corporation without complete BIA control. But the IRA also contained a clause authorizing the secretary of the interior to limit livestock on Indian reservations. Led by opposition leader Jacob Morgan of Shiprock, the Navajos voted against reorganization by a three-to-one margin. As a result, Collier reorganized the council system so that all resolutions required approval of the secretary of the interior.

The reservation was divided into land management districts, and individuals were issued grazing permits. By the end of World War II, Navajo livestock holdings were below permit levels, and Navajos could no longer rely on sheep as a primary source of income. Government intervention also changed the status of men and women. Agents, not understanding that among the Navajos property was held by women, refused to give them permits. This disenfranchised many women, creating problems within and between families, leaving some women in destitution, shifting sheep from one clan to another, and confusing the inheritance patterns. But the Navajos as a people resolved never to give up sheep, for as one Navajo has said, "When one sells all of his sheep, one will begin to feel useless. . . . When there are no sheep, one's home no longer is a real home" (quoted in L. Kelly 1968:163).

World War II and Beyond: Points of Contact

Like 40,000 other American Indians from almost every tribe, Navajos served with distinction during World War II in Europe and the Pacific. More than 3,400 Navajo men and women joined the U.S. Army and the Marine Corps and served in many capacities. Many played a very special and crucial role as Navajo Code Talkers. More than 400 Navajo men used a code based on Na-Diné, the Navajo language, that completely baffled the Japanese.

The Navajos performed their task so successfully that they were credited with being major influences in u.s. victories at Guadalcanal, Tarawa, Guam, and Iwo Jima.

Another 10,000 Navajos gained a reputation as outstandingly loyal and hard-working employees at shipyards, factories, and defense plants. Many went to California, both seasonally and permanently, beginning an urban migration pattern that still continues. The war also brought other changes, including increased contact with the outside world. Navajo veterans were convinced that the Navajos had to learn to be more effective in the non-Indian world if they were to survive. They led the Diné on a new journey, one that focused on education, business, and self-determination.

Given that the Diné could not return to a way of life that relied on sheep for its foundation, the tribe began to diversify and to make use of its other natural and human resources. The income from leases for the extraction of coal, oil, natural gas, uranium, and minerals has financed many tribal programs. Navajos have formed tribal utilities, a forest products program, and cooperatives, and have expanded agricultural production. Farmland acreage, particularly in irrigated plots, slowly increased so that a total of 34,800 acres was planted in 1980, and with the completion of a dam on the San Juan River in 1988, an additional 110,000 acres of irrigable land became available between Farmington and Shiprock.

Navajos began to develop new political skills aimed at expanding their self-government. The tribal council gradually took over responsibilities previously held by the BIA in water development, irrigation, emergency grain distribution, welfare and public works programs, trade licensing and fees, legislation and enforcement, and the regulation of domestic relations. In 1969 the Navajos changed their name to the Navajo Nation to emphasize their self-determinacy and independence and to remind Navajos and non-Navajos alike that they and their lands were separate and distinct from the states in which they were located. The tribe hired lawyers, who increased their presence in Washington, successfully lobbying Congress for funds to improve reservation roads and schools. The Navajos began to be effective in using the Anglo American court system as well. They argued before the u.s. Supreme Court in *Williams v. Lee* and secured a reaffirmation of Chief Justice John Marshall's 1832 decision that a treaty tribe was a domestic dependent nation and sovereign with respect to the states. This decision had far-reaching implications for many groups, denying states jurisdiction over civil and criminal cases involving Indians on reservations themselves.

The Navajo Nation established new educational programs, for the Diné believe that knowledge is power. By the late 1950s, Navajo children were attending more than forty-six elementary schools, which was no easy task in a society where people lived in small and isolated settlements scattered across an immense landscape. In the 1960s the Navajos developed contract schools, which took curriculum control away from BIA officials and placed it in the hands of local communities. The most famous, the Rough Rock Demonstration School, has served as a model for other schools by publishing educational materials in both Na-Diné and English. The goal of the contract schools is to combine traditional and Anglo American–style education and to teach students how to read and write in Na-Diné. The schools aim to make the concepts and skills that one needs to succeed in the outside world relevant while also emphasizing Navajo ways of learning through storytelling and observation. As Irene Stewart has noted, stories are "a source of a happy feeling, a quiet affirmation" of the Navajo Way (1980:48). The Diné Biolta, an educational association whose name translates as The People's Schools, also encourages Navajo parents to visit boarding schools and discuss Navajo history and culture. Strong women like Ruth Roessel have led in the development of curricular materials by many Navajo educators.

Higher education was also stressed following World War II. The Navajo Nation has set up a multimillion-dollar scholarship fund for qualified Navajo high school graduates to attend the colleges and universities of their choice. During 1980, for example, more than 1,900 Navajos received scholarships and attended more than 150 colleges in thirty states. For those who wish to remain on the reservation, Navajo Community College, which opened in 1969, offers associate degrees on several campuses.

Navajo workers at the Fairchild Electronics Company plant in Shiprock, New Mexico, in 1968. (Photograph by Helga Teiwes, Arizona State Museum)

Increasing contacts with universities in Arizona and New Mexico began in 1971 when the Navajo Division of Education was created. Its first goal was to work with universities to increase the number of Navajo teachers. Such collaborative efforts are very common today.

Even while they were adapting to new ways, however, Navajos did not abandon sheep raising as a way of life. Each family continued to have a small herd, and weaving, while not an economically viable occupation for many people, remains part of a very complex economic system. Chapter houses (grassroots local community centers where people discuss issues and grievances) have established economic development projects to combat poverty as part of emergency work-relief programs and to hire women for a month at a time to produce rugs. These women are paid a wage, and the chapter provides the raw materials and instruction in weaving. The completed rugs are sold at auctions around the reservation, the largest of which is held at Crownpoint. Weavers have also organized thriving cooperative associations, which give them control over the marketing process.

Walking in Two Beautiful Worlds

Today the Diné number more than 200,000. They constitute one of the most populous American Indian societies in the United States and live on a reservation about the size of West Virginia. In this respect, they have been highly successful, for Navajos consider large families a blessing. As in the past, the Beautiful Trail is filled with both blessings and trials. In today's world of computers, factories, mines, smog, fast food, obesity, and power plants, the Navajo people face the new challenge of preparing to meet a rapidly changing outside world on their own terms while preserving and teaching traditional cultural values. As in the past, this requires flexibility, for while the "gray gods"—old age, cold, poverty, hunger, sleep, lice, desire, and want—are still present, they come in new forms. While more Navajos own small businesses both on and off the reservation than ever before, and while the tribe has businesses and revenues totaling in the millions of dollars, the reservation also has one of the highest unemployment rates in the country. Per capita income is low, and poverty is widespread. To survive, many people must move away from their homes, relatives, and land. "A Navajo boy or girl wants a suitable home, a chance to live the life he has been taught, and an opportunity to find suitable work to support himself and later his family," Irene Stewart remarked (1980:33). More and more Navajos live in major cities throughout the West. Medical problems arise from today's less-active lifestyle, from junk food obtained at the trading posts, from a rising number of car accidents, and from a rising rate of alcoholism. Political corruption has marred recent administrations, while uranium mining and a falling timber market have created their own problems. It will take great wisdom to solve these problems while maintaining a harmonious world.

As if these issues were not enough, overgrazing remains a problem, and there have been conflicts over mineral and occupancy rights arising from the existence of joint-use areas with the Hopis and the Southern Paiutes. These problems date back to 1882 with the dividing of the two reservations. When the Hopi won exclusive occupancy rights in a court case in the late 1960s, the Navajos regarded this as a major defeat. The resulting program of relocating families is considered as devastating as the Long Walk, and many Navajos refer to it as the Second Long Walk.

Recently the Navajos have celebrated their history in a new way. In 1990 they asked their relatives from the north to meet them for a reunion at the Navajo Tribal Fair. Several members of the Sarsi tribe, an Athapaskan people from Alberta, Canada, came to Window Rock and participated in all aspects of the fair. The Navajos plan to invite other Athapaskan relatives to celebrate with them and renew their bonds, for they remember the important lessons they learned from the emergence story and from their clan migrations, that harmony, dedication, responsibility, and respect are important for survival.

THE YOEMEM (YAQUIS)

AN ENDURING PEOPLE

Thomas E. Sheridan

When anthropologists Edward and Rosamond Spicer first lived in the Yaqui Valley of Sonora in 1942, the terrible wars between the Yaqui Indians and the Mexican government were only a decade in the past. Military garrisons were still stationed in Yaqui pueblos, and many adults were just beginning to rebuild their lives in their ancestral homeland after years of exile. There were fields to be cleared, canals to be dug, crops to be planted. But the Yoemem were also reviving the elaborate religious and political systems that had bound them together for centuries in their Eight Sacred Pueblos: Ko'oko'im (Cocorit), Potam, Rahum, Torim, Vahkom (Bacum), Veenem (Belem), Vikam (Vicam), and Wivism (Huirivis).

A center of this revival was the western pueblo of Rahum, where the Spicers met a tall Yaqui elder named Juan Valenzuela. "Despite a very practical and direct approach and manner, he impressed me as living in some world which I knew very little about," Edward Spicer wrote. "He was a mystic. . . . His eyes were not bright and burning, but rather dim and burning. They were definitely looking off somewhere, clouded by the passage from our world to the other one" (Spicer 1992:114).

To revitalize Yaqui culture, Valenzuela had started a school in Rahum where he taught the myths of the Yoemem to younger Yaquis. Befriending the Spicers, he allowed them to transcribe and translate three of those texts—"The Flood," "The Singing of the Boundary," and "The Talking Stick and the Surem"—which Valenzuela himself had written down in a mixture of Yaqui and Spanish. The texts are sacred charters that establish the Yoe-

mem's God-given right to the Yaqui Valley. They speak of
Christian supernaturals like Saint Gabriel, Madam Virgin
Queen Guadalupe, Saint Michael, and Father God the Holy
Trinity. Yaqui place names intermingle with the Hebrew Sinai
and Golgotha. Catholic saints and symbols interpenetrate the
Yaqui landscape and infuse it with new meaning. The legacy of
Jesuit missionization in the 1600s and 1700s lives on in the
twentieth century.

But European patterns of belief did not replace Yaqui ones.
Rather, the Yaquis took what they had learned from the missionaries and made it uniquely Yaqui. In their theology, Jesus
Christ is a curer who preached and performed miracles in the
Hiakim, or Yaqui homeland. Pre-Christian deer dancers and
masked *pahkolam,* or old men of the fiesta (see pl. 8), perform
during ceremonies of Catholic origin. During the early 1600s,
the Yoemem defeated Spanish forces on at least three occasions
before allowing Jesuit missionaries to settle in their midst. For
the last 400 years they have zealously defended the Hiakim
from Spanish and later Mexican intrusion. In the process the
Yaquis have created a culture that is a vibrant fusion of the
European and the Native American. Centuries of struggle
against overwhelming odds have neither destroyed the Yaquis
nor forced them to assimilate into the mainstream of modern
Mexico. On the contrary, the crucible of conflict—economic,
military, and cultural—has only strengthened the Yoemem's
sense of themselves, where they came from, where they belong.
Yaquis like Juan Valenzuela have fought with rifles and they
have fought with words. According to Edward Spicer, they, like
the Jews, the Irish, the Basques, the Mayas, and many other
groups who have retained their language and defended their
sacred homeland in the face of intense persecution and repression, are an "enduring people" (Spicer 1980).

The Talking Tree

Nowhere is this abiding relationship with the past expressed
more clearly than in the myth of the Talking Tree. There are
many versions of the story among the Yoemem. The Yaqui Fel-

Yaqui maestro Juan Jesús Ujllolimea (center) and two Yaquis from Potam stand
beside three crosses marking the boundary between the Yaqui pueblos of Potam
and Vicam in Sonora. (Rosamond B. Spicer, early 1940s)

ipe Molina recorded this one when Luciano Velasquez told it at
Kompwertam (Compuertas), Sonora, in August 1982:

> In the time before the Spanish conquest the Surem lived in
> the area that is now west of Ciudad Obregón. Their river was
> called Yo Vatwe, Enchanted River. In this region the Surem
> had their homes, and they lived on both banks. Their houses
> were called hukim. They were built of sticks and mud and
> were about four or five feet high. They hunted, fished, and
> farmed to stay alive. Well, anyway, during those early times a
> tree was heard talking on a small hill called Tosai Bwia, White
> Earth. Some say it was heard at Omteme Kawi, but we say it
> was at Tosai Bwia. This tree was an old dead mesquite tree,
> and it made strange humming sounds. Nobody could under-
> stand the sounds. That bothered the Surem. All the intelligent

men in the Surem land were notified and told to visit the tree.
None of those intelligent men could figure it out. They all had
to admit failure. They could not decide what the meaning of
the tree was. These wise men knew of one other wise man
who lived near a little mountain called Asum Kawi, Grandmothers' Mountain. This man's name was Yomomoli, and he
had twin daughters. The wise men visited him and requested
that his two daughters interpret the talking tree. Yomomoli
told them that the girls didn't have a good vocabulary, so they
would be incapable of doing such a task. The men insisted.
Finally they convinced Yomomoli to take the girls to the talking tree. But Yomomoli did one thing first. He took his twin
daughters to the ocean. There they talked to a fish, so that
they might better understand the talking tree.

At Tosai Bwia the girls stood on either side of the tree, and
they began to interpret. The tree predicted Christianity and
baptism, wars, famine, floods, drought, new inventions, even
drug problems, and so on. After the tree had given all the
information it stopped making the sounds.

The Surem were happy, but they didn't really like some of
the things they heard, so they planned a big meeting. The
meeting was held near a water hole called Yo Va'am, Enchanted Waters. This is in the region between Vicam and the
modern town of Colonia Militar. There the Surem held both a
meeting and a dance of enchantment. At this meeting some of
the Surem decided to leave the Yaqui region, while others
decided to stay and to see these new things. At this dance of
enchantment, they say a real live deer came to dance for the
Surem. After the dance the Surem who were leaving cut up a
portion of the Yo Vatwe, wrapped it up in a bamboo mat, and
took it north to a land of many islands. Other Surem stayed
around and went into the ocean and underground into the
mountains. There in those places the Surem now exist as an
enchanted people. Those who stayed behind are now the modern Yaquis and they are called the Baptized Ones. (Evers and
Molina 1987:37–38)

To many Yoemem, the Surem are magical people who live in
the *huya ania*, the wilderness world. There all beings—plants,

animals, even rocks and springs—form one living community. The beings converse with one another, and the language they use is song. Through song they also make themselves intelligible to humans. Song is the universal tongue that unites people and animals. It unites the enchanted world of the Surem with the natural world of the Yaquis.

Deer Songs and the Flower World

The most ancient of Yaqui songs are the *maso bwikam,* or deer songs. They are the voice of *saila maso,* little brother deer. At nearly every major Yaqui ceremony, three men sing the songs while a deer dancer embodies the spirit of the deer itself, pantomiming its actions. The deer, like the Surem, is magical, a being from the *sea ania,* the flower world, where beauty and life abound (see pl. 6).

Flowers pervade Yaqui culture. Some Yaquis believe that the blood that flowed from the wound in Christ's side turned into flowers when it fell to the earth. Flowers symbolize grace from heaven (*teweka*), and therefore have Christian connotations. But flower symbolism is more ancient as well. Flowers also represent the sea ania, the flower world "beneath the dawn." All Yaqui deer songs describe that world, where both the archetypal deer and deer hunter live. Flowers are metaphors for all that is good and beautiful in Yaqui life, including the beauty of the deer and all the other beings who inhabit the magical pre-Christian world. When Wok Vake'o, the first Yaqui to compose deer songs, encountered a fawn, he sang:

> Little one, born in the night,
> Caressed by the fresh wind.
> Little one, born in the night,
> Caressed by the fresh wind.
> Where are you going then,
> Flower fawn among the flowers?
> Dressed in flowers, I am going.
>
> EVERS AND MOLINA 1987:50

The Struggle for Yaqui Land

Whenever deer songs are sung, they remind the Yaquis not only
of their magical past but also of the continued existence of enchanted worlds, which mirror the world in which the Yaquis live.
The Yoemem are therefore tied to the land by bonds that are
mystical as well as corporeal. According to another of the sacred
myths, four Yaqui prophets were joined by a band of angels in
1414. Together they walked the length and breadth of Yaqui
territory, singing its boundaries. Those boundaries were regarded as sacred. The land within them was given to the Yaquis
by God.

Edward Spicer, who devoted a lifetime of study to the Yaquis,
believed that the Singing of the Boundary was a myth that
developed in response to the bitter struggle for the Yaqui homeland during the nineteenth century. For most of the Spanish
colonial period, relations between the Yoemem and the Spaniards were peaceful. Most Yaquis lived within the Jesuit mission
system, cultivating their crops along the floodplain of the Yaqui
River. There they planted their seeds in the moist soil after the
floodwaters receded, just as they had done before the Jesuits
arrived. They also raised livestock given to them by the missionaries in the surrounding desert and mountains. The Jesuits introduced many changes into Yaqui society. Yoemem living in
eighty scattered settlements, or *rancherías,* were gathered into
eight mission communities—the Eight Sacred Pueblos of Yaqui
mythology. But the integrity of the Hiakim was preserved.

In 1740, however, the first major Yaqui revolt broke out. The
revolt was quickly suppressed, but violence flared once again
after Mexico gained independence from Spain in 1821. The
Mexican government believed that the progress of the Mexican
nation depended upon opening up such rich farmland to non-
Indian settlers, whom the Yaquis called *yoris.* Mexican attempts
to encroach upon sacred Yaqui territory met with fierce resistance. For the rest of the nineteenth century, conflict between
Yaquis and yoris gradually escalated into a war that eventually
assumed genocidal proportions.

The first wave of rebellion coalesced around a charismatic
leader named Juan Ignacio Jusacamea, who was also known as

THE YOREMEM (MAYOS)

Thomas E. Sheridan

When Spaniards pushed north up the west coast of Mexico in the sixteenth century, they encountered numerous groups of Native Americans who spoke Cáhitan languages within the great Uto-Aztecan language family. These groups—the Sinaloas, Tehuecos, Zuaques, Ahomes, Mayos, Conicaris, Tepahues, and Yaquis, among others —dominated northern Sinaloa and southern Sonora, occupying the rich alluvial floodplains of the Sinaloa, Fuerte, Mayo, and Yaqui river valleys. Today only two Cáhitan-speaking peoples remain: the Yoemem (Yaquis) of Sonora and Arizona, and the Yoremem (Mayos) of Sonora and Sinaloa (Sauer 1934; Spicer 1962).

Unlike the Yaquis, the Mayos were never able to wrest an indigenous reserve from the Mexican government. Instead, they live in villages or rancherías along the Río Mayo in southern Sonora, along the Río Fuerte in northern Sinaloa, and in the subtropical thorn forest between the two rivers. Many belong to ejidos (peasant corporate communities chartered by the Mexican government), which hold land and water in common for their members. Others work as wage laborers on private commercial farms (Crumrine 1977). They also raise small herds of sheep, cattle, and goats, and weave handsome woolen sashes and blankets. These textiles— along with baskets, earthenware ceramics, and wooden furniture—are sold for much-needed cash (Fontana, Faubert, and Burns 1977).

Like the Yaquis, the Mayos continue to carry out complex and elaborate ceremonies involving matačinim (crowned dancers), pariserom (Pharisees), čhapayékam (masked dancers), pahkoram (pascola dancers), and the maso (deer dancer) and maso bwikame (deer singers). Unlike that of the Yaquis, however, Mayo armed resistance to the Mexican state ended in the 1880s. Thereafter, Mayos responded to periods of great political and economic stress by organizing a number of religious revitalization movements led by Mayo prophets such as Teresa de Cabora. The daughter of a Mayo woman and a Mexican landowner, Teresa Urrea fell sick and "died" for twenty-four hours. When she recovered, she acquired great fame as a curer, and many Mayos revered her as a saint. Fearing an uprising, the Mexican government attacked one pasko (fiesta) where she was curing in 1892 and shot many of her followers. She and her father fled to Arizona, where people continued to seek her help (Crumrine 1977).

Another more recent revitalization movement was led by a young Mayo named Damian Bohoroqui, who claimed that "Our Father Old Man God" appeared to him in 1957. According to anthropologist N. Ross Crumrine, Damian's movement was profoundly conservative, reviving and elaborating upon paskos that had been part of Mayo cultural life since the Spanish colonial period. In a world where many Mayos are landless and live side-by-side with more powerful non-Mayos, religious revivals are a way of asserting Mayo values and Mayo identity. In Crumrine's words, "The Mayo prophetic movement and the whole process of cultural revitalization symbolize a people's effort to create their own destiny and to move, on their own terms, from a fluid liminal status to a more respected national status" (Crumrine 1977:156).

Cajeme (José María Leyva), the leader of the Yaqui resistance against the Mexican government during the 1880s. (Smithsonian Institution, National Museum of Natural History)

Juan Banderas. Banderas, which means flags in Spanish, received his name because he attempted to unite the Indians of northwestern Mexico under the flag of King Montezuma and the banner of Our Lady of Guadalupe. During the first stages of the revolt, he hoped to create an independent nation that included Mayos, Opatas, and Pimas as well as Yaquis. He soon replaced that ambitious vision with a more pragmatic attempt to win for the Yaquis political autonomy within the Mexican republic. The Mexican government opposed both goals, however, and in 1833 Banderas was executed in the Sonoran state capital of Arispe.

The next fifty years of struggle were characterized by outbursts of guerrilla warfare followed by brutal retaliation. In 1868, for example, soldiers commanded by General Jesús García

Morales took more than 400 Yaqui men, women, and children prisoner and locked them in the church at Bacum. When the prisoners tried to escape, the soldiers set fire to the church and turned their artillery on those who fled the burning building. By the time the carnage had ended, from 70 to 120 people had died.

Atrocities such as the Bacum massacre triggered the so-called Yaqui Wars of the 1880s. Led by José María Leyva, also known as Cajeme, the Yaquis organized themselves into formal military units and even constructed forts such as Buatachive in the Bacatete Mountains. Cajeme had risen to the rank of captain in the Mexican army during the struggle against the French in the 1860s. Drawing on that experience, he transformed Yaqui warriors into a formidable fighting force. In the end, however, the superior arms and numbers of the Mexican army overwhelmed Cajeme and his men. Cajeme himself was executed by firing squad in 1887. Thereafter, Yaqui guerrillas under Juan Maldonado, "Tetabiate," fought a valiant but losing effort against the tide of Mexican soldiers and *hacendados* (large landowners) sweeping across their land.

The Yaqui Diaspora

Dr. Manuel Balbás, a surgeon with the Mexican army that invaded Yaqui territory, recognized how deeply the Yaquis were attached to their land. In his words,

> The Yaqui alleges property rights in a belt of land, which includes the eight primitive towns of the river . . . and in addition the mountains of Bacatete which are on the North. . . . Outside this zone, which is very large and rich as a result of its excellent irrigable land, its woodlands and pastures, the Yaqui pretends absolutely no dominion, recognizes without argument private property rights and is submissive to all constituted authorities. But inside the zone . . . he recognizes no owners other than themselves, no authorities other than those named by themselves. That is to say that they desire to establish within this land an isolated fatherland from the rest of the world, absolutely free, independent and sovereign. (Spicer 1980:141)

Despite his awareness of Yaqui claims, however, Balbás was a typical intellectual of the late nineteenth century. Like the *científicos* who advised Mexican president Porfirio Díaz, he believed that "progress" was both inevitable and just. Expressing the widespread sentiment of the time, he wrote:

> The margins of the Yaqui river can constitute one of the greatest agricultural riches of the country. This river, like the Nile, inundates each year a great extension of land, leaving as it recedes to its bed, a magnificent accumulation of organic substances. . . . It cannot be permitted that such riches remain indefinitely in the power of men who do not know how or care to exploit them usefully. . . . Progress has great requirements and people who do not adjust to them must succumb, because the irresistible force of the universal action must push them ahead and absorb them in the course of its constant and powerful development. (Spicer 1980:141)

One particularly brutal course that "progress" took was the massive deportation of the Yoemem during the early 1900s. By the turn of the century, both the Mexican federal government and the state government of Sonora had concluded that a final solution to Yaqui resistance had to be developed. In the words of historian Evelyn Hu-DeHart:

> This final solution, which could have amounted to genocide were it not abruptly curtailed, was the massive and systematic deportation of all Yaquis out of Sonora. Most of the unfortunate Yaquis ended up in the notoriously brutal henequen plantations of Yucatán, where few survived the rigors of virtual slave labor. More than any military campaigns, deportation eventually broke down the spirit of the Yaquis, who could not sustain the resistance when they were physically dispersed and their families wantonly torn apart. (Hu-DeHart 1984:155)

Although the precise number of deported Yaquis has never been determined, Governor Rafael Izábal of Sonora bragged that he had shipped out two thousand by 1907. The following year the deportation program peaked, so the final figure must have been considerably higher.

Yaqui soldiers in about 1890. Yaqui guerrillas continued their struggle against Mexican occupation of the Hiakim, the sacred Yaqui homeland, until the Mexican Revolution broke out in 1910. (University of Arizona Library, Special Collections)

To escape the net, most Yaquis fled their homeland and fanned out across northwest Mexico. They worked in mines and haciendas, changing their names and masking their identities. Many Yaqui customs, especially public religious ceremonies, had to be abandoned. In the Sonora of Governor Izábal and the Mexico of Porfirio Díaz, being Yaqui meant peonage or death.

It was during these times of savage repression that many Yaquis crossed the border into the United States, especially Arizona. There they followed the tracks as itinerant railroad workers or labored in the fields around cities such as Yuma, Phoenix, and Tucson. They were political refugees who sought and found sanctuary hundreds of miles north of the Yaqui Valley. According to Spicer, the Yaquis "had become the most widely scattered native people in North America" (1980:158).

Yaqui prisoners in about 1900. To crush Yaqui resistance, the Mexican
government deported thousands of Yaquis to serve as laborers on haciendas in
Oaxaca and Yucatán. (Duane Bird Collection, University of Arizona Library, Special
Collections)

Waehma: The Christian Elements of Yaqui Religion

The deportation program ended with the Mexican Revolution in
1910. Many Yaquis participated in that struggle, forming all-
Yaqui battalions and fighting for leaders like Alvaro Obregón,
who was also from Sonora. Thousands of others filtered back to
the Yaqui Valley to put the shattered pieces of their lives back
together again. Sporadic conflict flared between the Yoemem
and the Mexican military for the next twenty years, and Yaqui
territory continued to be occupied. Nevertheless, the genocidal
onslaught against Yaqui society and culture had ceased.

One result was the gradual resumption of Yaqui religious
ceremonies. During the colonial period, Jesuit missionaries had

converted thousands of Yaquis to Catholicism. In the process, the Jesuits introduced numerous Catholic customs and celebrations practiced in seventeenth- and eighteenth-century Europe, including the elaborate public observances of Lent and Holy Week (Waehma, from *cuaresma*, the Spanish word for Lent). Such observances shriveled or died out in many areas of northwestern Mexico. Among groups like the Yaquis and the Tarahumaras, however, they survived and flourished, taking on a life and meaning of their own that bear little resemblance to orthodox Catholicism.

Today those celebrations bring together figures of Native American as well as European origin. The two major organizations responsible for Lenten ceremonies are known collectively as the Kohtumbre Ya'ura, or Customs Authority. One group— the Kabayum (horsemen; *caballeros* in Spanish)—represents the forces of good and serves as the guardian of Jesus. The other group, called by a variety of names, including Fariseos (Pharisees), represents the forces of evil. They are the ones who capture and kill Jesus. According to Spicer, the two organizations resemble "on the one hand, the Spanish *cofradía*, or secular religious brotherhood, and on the other the Western Pueblo ceremonial sodality" (1980:70).

But the rituals of Waehma also encompass beings that spring from ancient Yoemem traditions. Those beings—the deer dancer and the *pahkolam*—give Yaqui ceremonies a spiritual dimension that resonates at an older, deeper level. They bring the *huya ania*, the wilderness world, into the world of the pueblos.

Waehma culminates in the Looria Tenniwa (Running of the Gloria) on Holy Saturday morning, when the Fariseos storm the church in each Yaqui community and attempt to capture the image of Jesus. Jesus is defended by the Baptized Ones, especially the Kabayum and the Anhelwarda—the Angel Guard or Little Angels, consisting of small boys and girls. During this dramatic struggle, the Baptized Ones are joined by pahkolam and their singers along with the deer dancer and deer singers. To repel the invaders, the Baptized Ones hurl "flowers"—confetti, cottonwood leaves and flower petals—while the deer singers chant Looria Bwikam, or Gloria Songs (Evers and Molina 1987).

One of the Gloria Songs speaks of teweka, or heaven:

But there in heaven
   truly it is as if enchanted,
     my father.
But there in heaven
   truly it is as if enchanted,
     my father.

Over there in the flower-covered holy heaven,
   to see the glory,
     truly it is as if enchanted,
       my father.
But there in heaven
   truly it is as if enchanted,
     my father.

EVERS AND MOLINA 1987:56

Another of the Gloria Songs, on the other hand, refers to the huya ania itself:

Wilderness world,
   flower freely, is blowing,
     wilderness world.
Wilderness world,
   flower freely, is blowing,
     wilderness world.

Over there, I, in Yevuku Yoleme's
   flower-covered, enchanted,
     flower patio,
       flower freely, is blowing,
         wilderness world.
Wilderness world,
   flower freely, is blowing,
     wilderness world.

EVERS AND MOLINA 1987:58

Killing the Deer: Native American Elements of Yaqui Religion

Some observers are tempted to equate the deer dancer with Jesus himself, viewing both as sacrificial victims. But such a view places too much emphasis on the Christian aspects of Yaqui ritual and belief. The religion of the Yoemem is not merely a variant of Latin American folk Catholicism. On the contrary, it represents a complex tapestry of both aboriginal and Christian symbols and supernaturals. Spicer called it "a conjunction of world views" (1980:62). In his opinion, it is a new religion, one that draws from both European and Native American traditions.

In contrast to the ceremonies of Waehma, perhaps the fullest expression of indigenous Yaqui ritual is another drama of death and renewal: the *maso me'ewa,* or killing the deer. The killing-the-deer ceremony takes place on the first anniversary of a relative's death. It marks the end of a year of mourning by the kin of the deceased and releases the spirit of the dead person from its last year of confinement on "this weeping earth."

The anniversary celebration itself is known as *lutu pahko, pahko* being a general term for any Yaqui ceremonial occasion. Like any other pahko, the deer dancer and deer singers are joined by the *pahkolam,* the "old men of the pahko," who serve as hosts. Next to the deer dancer, the pahkolam are the most distinctive Yaqui ceremonial figures. Stripped to the waist and wearing small wooden masks, the pahkolam mime, dance, and joke with each other and the crowd. Their clowning serves as a counter-point to the dignified demeanor of the deer dancer himself. Yaquis call their antics *yeuwame,* or games. The games form a hilarious and often pointed commentary not only about Yaqui society but also about society at large.

Despite their spontaneity, however, the pahkolam perform an established sequence of yeuwame at the killing-the-deer cere-mony. First they mimic Yaqui farmers pestered by a raccoon. Then they act out the pursuit of the deer by mountain lions. Finally, they chase and kill the deer themselves. Occasionally they also depict the arrival of the rains. The last three games are always accompanied by deer songs.

The culmination of the maso me'ewa, of course, is the stalking of the deer. One pahkola becomes a hunting dog, while the other three act as hunters. They request the use of the deer singers' instruments, holding them upside down, jumbling words and rhythms. Their burlesque stands in stark contrast to the poi-gnancy of the hunt itself, which is communicated through the

YAQUI AND MAYO PAHKOLA/PAHKORA MASKS

Thomas M. Kolaz

Perhaps no figures are as distinctive at a Yaqui ceremony (a *pahko*) as the *pahkolam,* or old men of the fiesta. Representing beings from the *huya ania,* the wilderness world, the pahkolam serve as ceremonial hosts throughout the pahko. They also dance, clown, mime, and joke with the crowd and each other. No pahko can begin or end without their presence. With cocoon rattles wrapped around their legs and masks on their faces (see pl. 8), the pahkolam make the wilderness world palpable for Yaquis at ceremonies throughout the year.

The masks themselves are relatively small and cover only the pahkola's face. Carved from soft wood such as cottonwood, often by the pahkolam themselves, the masks are traditionally painted black with red and white design elements—triangles, insects, animals, plants, for example—along the sides and top. There is always a cross on the forehead and sometimes on the chin as well. Above the eyes and under the mouth, the mask also bears tufts of hair taken from the tails of goats, horses, or cows. Human and goat face masks are the most common, but Frank Martínez, a pahkola and carver from Old Pascua in Tucson, carves coyotes, lions, and roosters as well. Pahkolam dance with their masks resting on the back or side of their head while one musician plays the harp and one or more musicians play the violin. They dance with the masks over their faces when the *tampaleo,* or drummer, plays the drum and flute.

The Mayos, linguistic relatives of the Yaquis who live in southern Sonora and northern Sinaloa, also have old men of the fiesta, whom they call *pahkoram.* Mayo masks fall into two styles: Sonoran and Sinaloan (see pl. 7). The Sonora-style masks are larger than either Yaqui or Sinaloa-style masks, and their base coats and design elements come in many more colors. White goat hair often completely encircles the masks. Sinaloa-style masks, in contrast, are the same size as Yaqui masks—about five inches by eight inches—but they are unpainted or lightly stained and have very few design elements. The Sinaloa-style carvers also produce more goat than human face masks.

From simple pieces of wood, Yaqui and Mayo Indians have carved out a place within the American Indian masking tradition. The small painted wooden masks help transform men and boys into ritual hosts, reminding the Yoemem of the enchanted worlds that surround and permeate the world of the *pweplum,* or town.

actions of the deer dancer and the words and music of the deer singers. Like so many other deer songs, the songs of the killing-the-deer ceremony are filled with images of flowers even when they talk about death.

Toward a place where
I could not find safety I went.

Although unseen in the wilderness
I am just running,
My antler crown with these three branches
is showing, moving.
. . . . .

Flower-covered grove, as I am walking to you,
I am talking to you, flower-covered grove.
. . . . .

Not wanting to die,
dodging through the wilderness.
. . . . .

with my head hanging down
toward the ground,
as I am walking,
with foam
around my mouth
as I am walking,
. . . . .

Never again I,
will I on this world,
I, around will I be walking.
. . . . .

Killed and taken, killed and taken,
there in the wilderness,
I am killed and taken.

EVERS AND MOLINA 1987:159–68

After the deer has died, the deer dancer is laid on a bier of branches, where the pahkolam "skin" and "butcher" him. But then, miraculously, the deer is transmuted into a "flower" and reborn in the wilderness world:

Oh, put a flower on me
    from flower-covered person's flower body.
Put a flower on me
    from flower-covered person's flower body.
Over there, I
    in the flower-covered flower opening,
        as I am standing,
    covered with dust,
        as I am standing,
    covered with mist,
        as I am standing,
Put a flower on me
    from flower-covered person's flower body.

EVERS AND MOLINA 1987:170

This magical transformation takes place even as the pahkolam roast the meat of the deer and throw his intestines to the vultures. While they are doing so, the deer singers repeat:

My enchanted flower body,
    fire, above the fire,
        side by side is hung.
My enchanted flower body,
    fire, above the fire,
        side by side is hung.

Over there, I, in Yevuku Yoleme's
    flower-covered, flower patio,
        here I am scattered,
            I become enchanted,
        here I am scattered,
            I become flower.
My enchanted flower body,
    fire, above the fire,
        side by side is hung.
· · · · ·
Over there, I, in Yevuku Yoleme's
    flower-covered, flower patio,
        I am just glistening,
            sitting out there,

> here I am scattered,
>   I become enchanted.
> My enchanted flower body is glistening,
>   sitting out there.
>
> EVERS AND MOLINA 1987:171–72

The ritual of the killing of the deer therefore sets the grim reality of the hunt against the transcendent spirituality of the flower world and the huya ania. The wilderness world, immanent and ethereal, survives and triumphs in the face of blood, death, and the absurd antics of the pahkolam. Such a message is far older than Christianity in the Sonoran Desert, and it is expressed through a ceremony that is full of creative and unexpected juxtapositions. In the words of Felipe Molina, a Yoeme from Arizona who is a deer singer himself:

> Everything the deer dancer uses in his dance has held life. The cocoon rattles around his legs were once homes of the butterflies. As we dance we want the butterfly to know that, even if he is dead, his spirit is alive and his house is occupied. The gourd rattles in the dancer's hands give life to the plant world. The rattles around the dancer's waist are deer hooves. They represent the millions of deer who have died so that men might live. (Evers and Molina 1987:129)

Yaqui Ritual in the Modern World

Catholic ceremonies and supernaturals have enriched Yaqui religion, but they have not encompassed it. According to Larry Evers and Felipe Molina, who collaborated on the award-winning *Yaqui Deer Songs/Maso Bwikam*, "after more than three and one-half centuries of dialogue with Catholicism, not to mention various other versions of Christianity in this century, Yaquis continue to hold the figures of Christ and the *maas* [deer dancer] explicitly apart in their ceremonies" (Evers and Molina 1987:130).

Yaqui public rituals also continue to be the single most important expression of Yaqui ethnic identity in the modern world. The rituals are universal in the sense that they are open to outsiders and seek blessings for all men and women, not just

Miguel Romero, a Yaqui elder and head dancer (*monarca*) of the *matachin* society, carries a statue of the matachines' patroness, the Virgin Mary, in a cloth-wrapped box in a ceremony at Potam, Sonora (Rosamond B. Spicer, early 1940s).

Yaquis. Nevertheless, they clearly set the Yaquis apart from their neighbors—from Mexicans in Sonora and from Mexican Americans, Anglos, and O'odham in the southwestern United States. The Christian overtones are tantalizing in their similarity to folk Catholic practices among other groups. Figures like the deer dancer and the pahkolam, on the other hand, make audiences dramatically aware of the differences between Yaqui culture and their own. And even Yoemem who don't participate as *chapayekam* (masked dancers; literally, "those with long noses"), *matachinim* (crowned dancers accompanied by violins and guitars), *pahkolam, Kabayum,* or in any of the other organizations that carry out Yaqui ceremonies, share a sense of distinctiveness that is reinforced each time the deer songs are sung or the Baptized Ones repel the Pharisees during the Running of the Gloria on Holy Saturday.

Such rituals are one of the ways in which the Yaquis have survived persecution and diaspora to become one of the enduring peoples of the world. They provided a vivid focus for Yaqui society during the colonial period when Jesuit missionaries lived and worked among them; they knit people together in the Eight Sacred Pueblos during the violent conflicts of the nineteenth century; and they helped Yoemem in the refugee communities of Arizona recreate a vital social and cultural life of their own at a time when families had been torn apart and dispersed from Yucatán to California.

Today, the Yaqui struggle to survive as a people takes different forms. Pitched battles and bloody guerrilla warfare no longer rage across the Yaqui Valley. Nevertheless, conflict over the Hiakim persists. In 1937, Mexican president Lázaro Cárdenas issued a decree that stated:

> To the Yaqui tribe is ceded the whole extension of workable land located on the right bank of the Yaqui river, with the necessary water for irrigation from the dam in construction at Angostura, similarly all the mountain lands known as "Sierra del Yaqui" [the Bacatete Mountains], for which areas shall be provided the resources and elements necessary for the better utilization of their lands. (Spicer 1980:263)

Even though the area within the Yaqui Indigenous Zone only encompassed about a third of the Hiakim, Cárdenas' decree was unprecedented in Mexican history. For the first time the Mexican government recognized the legitimacy of Yaqui claims and pledged to protect them. For the first time since the Jesuit mission period, land in the lower Yaqui Valley was reserved for the exclusive use of Yaqui farmers and ranchers. The government also promised to build an enormous dam and canal system to bring water to irrigate the land. It took more than thirty years to

complete that system, but by the 1970s the Yoemem were cultivating thousands of acres of the best farmland in Mexico.

During the same period, however, non-Yaqui farmers brought
nearly 500,000 acres of irrigated land under the plow. A city of
half a million people—Ciudad Obregón—arose on the southern
plains of the Yaqui homeland. The Yaqui Valley became the
cutting edge of industrial agriculture in Mexico, so productive
that Norman Borlaug and the Rockefeller Foundation established the wheat-breeding station there that triggered the international Green Revolution.

In the process, thousands of landless peasants and wealthy
agribusinessmen have crowded around the margins of the Yaqui
Indigenous Zone, threatening its boundaries and illegally rent-
ing Yaqui land. The battles are fought in the offices of the Mexican government now, where Yaquis struggle to keep their relentless neighbors at bay. Felipe Molina quotes a letter written
by Yaqui elders in 1991. "We belong to a people of indomitable
fighting spirit," the letter begins. "Our race never gave in during
the dark days of the Spanish and we set an example as a people
who never allowed ourselves to be conquered, nor for our lands
to be taken away during the colonial period." The writer then
talks about how Cárdenas recognized the Yaquis' title to their
land: "With the strength which has been recognized as ours, and
the great fighting and working spirit of our people, the Yaqui
community began to be a prosperous and united people. However, our richness was coveted; the renters came, the profiteers,
the great landholders, who, with the help of evil officials, grew
rich at the expense of our people" (Evers and Molina 1992:19).

When Yoemem in Sonora discuss the myth of the Talking
Tree today, they say it foretold the coming of an enormous black
snake, which they equate with Mexican Highway 15. The highway brings not only death and destruction in the form of accidents but also the assault on Yaqui society and culture that
threatens to erode the old traditions and the sacred homeland.
One of the ways the Sonoran Yaquis are attempting to resist this
trend is by taking control of the government-funded development programs designed to increase agricultural production on
Yaqui lands and diversify the economic base of Yaqui commu-

Yaqui drummer Juan Tampaleo (*tampaleo* is the Yaqui term for a drummer) and his grandchildren at his home in Potam, Sonora. (Photograph by David Burckhalter, 1989)

nities. The most ambitious is the Programa de Atención Técnica Integral para las Comunidades Yaquis (PATICY). Yoemem now determine the goals and implementation of PATICY, including a greater emphasis on traditional subsistence crops such as corn and beans and less emphasis on commercial crops like wheat and soybeans, which dominate the fields of the Yaqui Valley. Like many such community development projects, however, PATICY has factionalized several of the largest and most important Yaqui pueblos, particularly Potam, Vicam, and Rahum. New sources of money and power have challenged the authority of traditional Yaqui pueblo governors.

The struggles of the Yoemem in the United States are more diffuse. Most Yaquis in Sonora are farmers or stock raisers. In Arizona, Yaquis make their living as wage earners, often in low-paying jobs when jobs are available at all. Unemployment reaches as high as 50 percent in some communities. The first wave of Yaqui immigrants worked on the railroad. Then many of them became farm laborers when Arizona's cotton boom began during World War I. Today most live in Tucson or the cities of the Salt River Valley, where they work as construction workers, gardeners, secretaries, teachers, health care professionals, and administrators. Conflict does not erupt over agricultural credit or agrarian reform, but tensions exist between the traditional demands of family and community and the modern demands of job, school, and the relentless mobility of the Sunbelt.

One group of Yaquis in Tucson met the challenges of poverty and poor housing by forming the Pascua Yaqui Association, which received 200 acres of land southwest of the city in 1964. Two years later, a grant from the Office of Economic Opportunity enabled the association to begin building New Pascua Pueblo, a planned community, on the land. The people of New Pascua won recognition as a U.S. Indian tribe in 1978, allowing them a degree of political autonomy as well as access to resources from the Bureau of Indian Affairs (BIA). In contrast, the older Yoemem communities of Pascua and Barrio Libre in Tucson, Yoem Pueblo in Marana, and Guadalupe on the outskirts of Tempe remained Yaqui enclaves unrecognized by the BIA. They had to pursue other federal grants to get the funds they needed to better their homes, their schools, and their jobs.

It is testimony, then, to the vitality of Yaqui society that deer songs and other cultural expressions are experiencing a renaissance on both sides of the border. Young Yoemem from Pascua, Marana, and Guadalupe are literally sitting at the feet of master deer dancers and deer singers from Sonora. Groups like the Coyote Society have been revived. In the face of drugs, gangs, television, and upward mobility, many Yoemem continue to find strength in tradition. The huya ania illuminates the world of tractors, irrigation canals, freeways, and nine-to-five jobs. The spirit of *saila maso*, little brother deer, thrives.

Plate 1. Shell and turquoise necklaces worn by Navajos. Such necklaces are manufactured at New Mexico pueblos. The necklace at right was made before 1925; the necklace in the center before 1970; the necklace at left before 1937. (Photograph by Helga Teiwes, Arizona State Museum)

Plate 3. Navajo squash-blossom necklaces. The necklace at the left is from around 1900, that in the center is from the early 1920s, and the necklace at the right was made around 1930. Wearing silver and turquoise jewelry is a sign of wealth and blessing. (Photograph by Helga Teiwes, Arizona State Museum)

Plate 2. Navajo woman's dress made in about 1865 from hand-spun and raveled yarns. Based on the Pueblo woman's manta (dress), the Navajo dress consisted of two rectangular pieces of woven wool, sewn down the sides and along the top, with slits left for the head and arms. The designs on the top and bottom borders replicate that on the so-called Navajo wedding basket. (Photograph by Helga Teiwes, Arizona State Museum)

Plate 4. A Navajo Ganado-style rug with Germantown and raveled red yarn and a cotton warp from about 1890. Trader Lorenzo Hubbell opened a trading post at Ganado, Arizona, in 1878. He encouraged weavers to use bold reds and to incorporate old styles and motifs into their rugs. (Photograph by Ken Matesich, Arizona State Museum)

Plate 5. A Navajo Wide Ruins rug, made with hand-spun vegetal dyed wool in 1952. The trading post at Wide Ruins, south of Ganado, was a center for experimentation in the use of natural dyes beginning in the 1930s. (Photograph by Ken Matesich, Arizona State Museum)

Plate 6. The *sea ania* (the Yaqui flower world), where *saila maso* (little brother deer) resides. (Mural by Danny Leon, Old Pascua, Tucson, 1993, commissioned for the Paths of Life exhibit; photograph by Ken Matesich, Arizona State Museum)

Plate 7. Mayo *pahkora* masks. The Mayos—linguistic and cultural relatives of the Yaquis—live in southern Sonora and northern Sinaloa. Like the Yoemem, Mayo ceremonies represent a fusion of Christian and Cáhitan supernatural figures and beliefs. The mask at center was made by Alcario Buitimea from the Río Fuerte in Sinaloa, around 1970; the mask at left dates to around 1930 and is from Huatabampo, Sonora; and the mask at right was made around 1980 along the Río Mayo in Sonora. (Photograph by Helga Teiwes, Arizona State Museum)

Plate 8. Yaqui *pahkola* masks and a disk rattle (*senasum*). The *pahkolam,* or "old men of the fiesta," perform at many Yaqui ceremonies, including the *maso me'ewa* (killing the deer) ceremony. The rattle was made by Ramón Duarte around 1940. The mask at right, from Old Pascua in Tucson, dates to 1932, and the mask at left was made in Potam, Sonora, before 1942. (Photograph by Helga Teiwes, Arizona State Museum)

Plate 9. *Apache Creation Story: The Things Legends Are Made Of.* Duke Wassaja Sine (1955–   ), 1988. Pen and ink and watercolors. (Painting in the ASM collections; photograph by Helga Teiwes, Arizona State Museum)

Plate 10. Western Apache coiled basketry: a jar, made between 1890 and 1915; a bowl, made around 1900; and basketry materials consisting of, from top, sumac twigs (*Rhus trilobata*), yucca root (*Yucca baccata*), and martynia (*Proboscidea parviflora*). (Photograph by Helga Telwes, Arizona State Museum)

Plate 11. A San Carlos Apache agave wood fiddle, made by Amos Gustina before 1925. (Photograph by Helga Teiwes, Arizona State Museum)

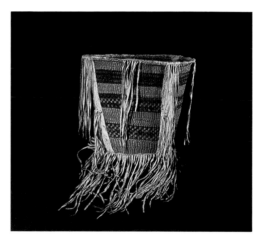

Plate 13. A Western Apache painted twined burden basket from Carrizo, Arizona, made before 1923. (Photograph by Helga Teiwes, Arizona State Museum)

Plate 12. A Western Apache pitched, twined bilobe canteen, made around 1930. Apaches used bilobe-shaped bottles as canteens. A rope cinched around the middle served as a carrying strap, and the opening at the top was plugged with bark or a bundle of grass. (Photograph by Helga Teiwes, Arizona State Museum)

Plate 14. Western Apache beaded buckskin moccasins, made before 1935. (Photograph by Helga Teiwes, Arizona State Museum)

Plate 15. A Western Apache war shield, made in the 1800s. (Photograph by Helga Teiwes, Arizona State Museum)

Plate 16. Canyon Creek in the Grand Canyon in the late 1850s. The Ives expedition encountered the Hualapais here. (Illustration by Heinrich Balduin Möllhausen, Museum für Volkerkunde, Berlin)

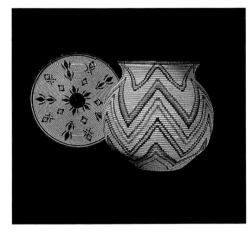

Plate 17. A Havasupai basketry bowl and jar, both made before 1924. (Photograph by Helga Teiwes, Arizona State Museum)

Plate 18. A Hualapai twined jar, at left, made before 1911 and a "fruit basket" made before 1959. (Photograph by Helga Teiwes, Arizona State Museum)

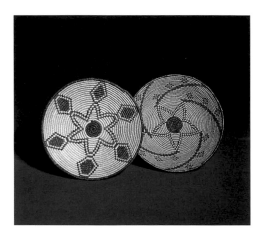

Plate 19. Two Yavapai coiled baskets. The one at left was made around 1900, and that at right was made at Camp Verde before 1923. (Photograph by Helga Teiwes, Arizona State Museum)

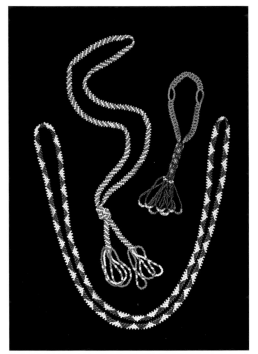

Plate 20. Hualapai beadwork. At top left is a necklace by Gloria Kausser, at top right is a belt tassel by Tonita Paya, and on the bottom is a necklace by Gloria Kausser. All were made before 1972. (Photograph by Helga Teiwes, Arizona State Museum)

THE INDÉ
(WESTERN APACHES)
THE PEOPLE OF THE
MOUNTAINS

Bruce E. Hilpert

On a bright day in June, Edgar Perry, the director of the White Mountain Apache Culture Center, leads a small group of his staff up the long trail to the 11,590-foot peak Dzil Ligai, or Mount Baldy. Along the way they enjoy the 100-mile view across almost half of Arizona and collect trash left by careless hikers. Near the summit, a sign informs non-Apaches that they may venture no farther—for this is a special place, a mountain sacred to the Apache people. When the group reaches the peak, they stop to say prayers of thanks for their mountain homeland.

Each year, countless members of the White Mountain Tribe make this journey to seek spiritual renewal from the sacred mountain and to reaffirm the Apache way of life. The Western Apache people have attachments to their mountain homelands that go far beyond a love for the beautiful landscape of central Arizona. The mountains not only provide a wealth of economic resources to support Indé timber, cattle, and recreation industries, they are also tied spiritually to the preservation of Apachean culture and the well-being of the people. In fact, through traditional stories, Western Apache values are embodied in the mountains themselves. As Annie Peaches told anthropologist Keith Basso, "The land is always stalking people. The land makes people live right. The land looks after us" (Basso 1987:95).

Origins: The Mountain Spirit People

While some Western Apaches accept anthropologists' ideas about linguistic ties to Athapaskan speakers far to the north, traditional creation beliefs are rooted firmly in the mountains of the Southwest. Edgar Perry's annual visit to Dzil Ligai puts him in touch with the forces of creation that are still present in the mountains of the Apache people.

There are several recorded versions of Western Apache creation stories. Most begin with the Creator, or Sun, making the earth, which was first inhabited by animals, monsters, and giants. In caves beneath the mountains lived the Gaan, the Mountain Spirit People—supernatural beings who are the protectors, teachers, and role models for the Apaches.

In their spirit world under the mountains, the Gaan live the Apache way, hunting, gathering, and farming. After the Creator made the first Apache people, the Gaan taught them to perform ceremonies that gave the Indé spiritual power. The Gaan watch over the Apache people, protecting them, admonishing them, and helping them to live properly. In effect, the Gaan are the spiritual ancestors of the Western Apaches and embody the essence of Apache culture and virtue. At times they leave their homes in the sacred caves of the mountains, appearing to the people to impart their power for healing, to ensure the well-being of the people, or to steer them back to the proper path when necessary. At ceremonies and other important events, they appear as dancers, sometimes called Crown Dancers. These dancers always appear in a group of five: four Gaan representing each of the sacred directions and a messenger who communicates with them. Today, the Gaan dancers still appear at puberty ceremonies, healing ceremonies, and other public events, reminding Apaches of their spiritual ties to the Mountain Spirit People.

One of the most popular traditional stories tells of the origin of the Gaan Dance, when the Apache people first learned to tap their power. Like other traditional tales, it is still told by elders to their children and grandchildren but only during the winter and only at night. Among the Western Apaches, each group or band may have its own version, with significant place names that tie the story to its own local area and a specific cave. The following summary is based on the story as told by Rudolph Kane from Cedar Creek on the Fort Apache Reservation (Evers 1978:4–8).

At one time, long ago, the Apache people had strayed from the proper way of living taught to them by the Creator. They were not living up to their responsibilities to care for their land and the things that live there. The Creator sent the Gaan to remind the people of how to live the right way. The people improved for a while, but, being human, they soon returned to their old ways. As a result of their actions, they found that there were no more deer in the mountains to feed the people. Desperate times had fallen on the Apaches.

One day, a young boy and his spotted dog were out in the woods hunting deer when they discovered a cave high up in the mountains. As they explored the entrance to the cave, the boy heard singing and the loud whirring of bull-roarers. When he crept closer to the sound, he discovered a group of Gaan dancing in their underground home.

That night the spotted dog returned to camp without the boy. His mother and his relatives went out to look for him, but to no avail. The dog led them to the cave, and the people suspected that he had gone to the home of the Gaan. The elders of the different camps got together and decided that the people must not have been living right. That night the people gathered to sing and dance so that the Gaan would come to them.

Four Gaan came to them: the black from the east, the blue-green from the south, the yellow from the west, and the white from the north. The people could sense the power of their presence. As these Mountain Spirits danced, the spotted dog ran from the crowd of people, barking and jumping up on one of the Gaan. The boy's mother realized that it was her son. As he struggled to remove his mask, he shouted, "It's no use. This mask, I can't get it off!" He had become one of the Spirit People.

All his relatives were sad, but he told them, "Wherever you go, pray. And I will pray for you, too." And that is how a man turned into a Gaan and how the Mountain Spirit People imparted their power to the Apache people.

Today, the power and knowledge of the Gaan are closely associated with the mountains. Some Apache elders still know the location of the cave that is mentioned in the story. But most important, the Mountain Spirits still embody the values and beliefs that make the people Apache.

APACHE CREATION STORY: THE THINGS LEGENDS ARE MADE OF

Bruce E. Hilpert

(SEE PLATE 9)

Duke Sine's paintings interpret the ceremonies, history, and dreams of the Apache people. Born in San Carlos, Arizona, he studied at the Institute of American Indian Art in Santa Fe. His works are in the collections of the Heard Museum in Phoenix and the Arizona State Museum in Tucson. In 1988 Duke Sine described his painting *Apache Creation Story: The Things Legends Are Made Of*, which appeared in the Arizona State Museum exhibit Among the Western Apache, in the following manner:

It all begins with the Creator, the Giver of Life. He made the earth, but there were no people there. There were just animals and monsters, but no humans. Changing Woman represents the first woman, or Mother of the Apache people. Changing Woman was lonely; she was the only one on earth. The Creator decided that it was time for people to populate the world.

Inside the earth during this time, there were the Mountain People, supernatural people who were the ancestors of the Apache. They all had different powers: some were medicine men, some were hunters, some were warriors; all the things that Apache people are today. These people are the ones that the Gaan Dancers impersonate. They are the real Gaan. They dress like regular people, but they are supernatural. During this time, the world was barren.

Changing Woman was kneeling and praying, looking directly into the sun and she was impregnated by the Creator. At this time, she was struck by lightning four times. After that, Child of the Water was born. Changing Woman had children at different times, but they always got eaten by monsters, the beings that ruled the world. Apache people believe that they came from within the earth. In this painting you see Child of the Water coming through the water, or through the surface of the earth.

There was another person on the earth; this was Slayer of Enemies. He was a warrior that represents the warrior spirit in the Apache people. Together, Child of the Water and Slayer of Enemies battled with the monsters, or the beings that ruled the earth, and one by one they defeated them. There were eagles, a giant, and different animals like buffalo and antelope that had to be defeated so that the world could be populated. After they slay the monsters, the world was ready for the human race.

To the right are the game animals that were imprisoned underground. After Slayer of Enemies and Child of the Water defeated these beings, they let the game animals loose to help man survive. Each animal has its own spirit and its own positive power that is beneficial to mankind.

After they did this, Changing Woman, Child of the Water, and Slayer of Enemies taught the Apache people how to do the Changing Woman Ceremony—the Sunrise Ceremony—and how the Gaan can help to heal, and protect, and take care of the Apache people in the way that the Creator intended

them to be. The Gaan are very powerful beings and are the mainstream of Apache religion.

In the bottom of the painting is the dawn
because this is the dawn of mankind. The
hummingbirds represent the directions and
have a color for the directions. They are
going out to all corners of the world telling
the world and the universe that the Apache
people have emerged.

The other side of the painting has to do
with the legend of Coyote, the Trickster, who
represents the negative things that man does.
One time there was a bag that the Creator
told them not to mess around with. Being
Coyote, he couldn't help but do the wrong
thing. He grabbed the bag and escaped with
the fire. If he hadn't done this, there would
have been daylight forever; but since he stole
the fire, there was night, and with it came the
other spirits of the night. On this side are the
snake, the owl, the bear, and the bat. The
Apache people don't eat them or touch them
or have anything to do with them because
they have a negative power that can hurt you.

Many Closely Related Bands

The Apaches of Arizona are probably the best known but least understood of the Indians of the Southwest. Through hundreds of western movies and novels, there is scarcely a person alive who is not familiar with the stereotype of the Apache warrior. Yet, due to these pervasive and inaccurate images, most people know little of the truth about the Apache people and their cultures.

The term *Western Apaches* is used by anthropologists to refer to the San Carlos, White Mountain, Cibecue, Northern Tonto, and Southern Tonto Apaches living in the mountainous regions of eastern Arizona. The Western Apaches share a language and many aspects of their culture with other Apachean groups of the Southwest and the Plains—the Chiricahua, Lipan, Jicarilla, Mescalero, and Kiowa-Apaches, and the Navajos. While recognizing cultural affinities with other Apacheans, the five distinct groups of the Western Apaches have sensed a certain cultural and linguistic unity that sets them apart from the Chiricahua Apaches and the eastern groups (Goodwin 1969:1).

Together with the Navajos, the five Western Apache groups belong to a linguistic family called the Southern Athapaskan. Their linguistic relationship to Northern Athapaskan speakers of western Canada and Alaska was first recognized by William W. Turner in 1852. For decades, linguists have investigated this intriguing connection. By the 1930s, anthropologists had concluded that the Southern Athapaskans left their northern homeland, traveling more than two thousand miles in a migration that may have lasted many generations. While the actual route may never be known, many scholars today favor the theory that mobile groups of hunters followed herds of bison south through the High Plains. Glottochronological studies, using divergence in language to date the split between related groups, indicate that the Southern Athapaskan migrations took place no more than a thousand years ago, with entry by successive Apachean groups into the Southwest at about the time of the earliest Spanish contact in the mid-1500s.

When Francisco Vásquez de Coronado first passed through the mountains of eastern Arizona in 1540, he described the re-

Dutchy (left), Alchesay (right), and General George Crook in 1885. Although
they faced overwhelming odds, the White Mountain, San Carlos, Tonto, and
other Western Apaches fought General Crook's soldiers. They finally
surrendered in 1873 and later served as scouts against the Chiricahua Apaches.
Chief Alchesay received the Congressional Medal of Honor. (Gatewood
Collection, Arizona Historical Society)

gion as a *despoblado,* an uninhabited area. While some scholars
believe that the Apaches living in the rugged canyons simply did
not want to be seen, they probably had not yet expanded that far
west. Indeed, even the first reference to the Apachean-speaking
Querechos near the Acoma pueblo in western New Mexico by
Antonio de Espejo and Gaspar de Luxan in 1583 does not confirm the presence of Western Apaches in present-day Arizona,
but it does suggest that their travels to hunt and gather foods
had led them into the Arizona mountains. By 1700 the Western
Apaches had firm control of the region south of the Mogollon
Rim, moving in from the north, as recounted in numerous clan
legends.

Each Western Apache group had its own subsistence pattern,
based on individual preferences and the resources available in its
own locale. The four territorial bands of the San Carlos group

lived mostly in the rolling high desert surrounding the Pinal,
Apache, Mescal, and Catalina mountain ranges. The Eastern and
Western bands of the White Mountain group covered a vast
territory from the Pinaleno Mountains in the south to the high
plateau north and east of the White Mountains. The Cibecue,
Southern Tonto, and Northern Tonto groups utilized the high
country from the Salt River north to present-day Flagstaff.

The bands within each of the five groups occupied their own
territories but shared resources and recognized ties to the other
bands of the group. However, the bands did not owe political
allegiance to the larger group, the hallmark of a "tribe." Each
band acknowledged a leader chosen by consensus for his experience, wisdom, and demonstrated good judgment. Band members

Ration lines at the San Carlos subagency in about 1880. Reservation life was especially hard at San Carlos, where even firewood had to be rationed. Groups often left the reservation just to gather food. (Arizona Historical Society)

traced kinship through matrilineal clans. They married outside their clan but usually within their larger group. When a man married outside his group and went to live with his wife's family (a matrilocal system), he was destined to spend the rest of his life as an outsider, a nonrelative, in his own home.

The Apache wars of the 1870s and 1880s disrupted the lifestyle, the territories, and the organization of the five Western Apache groups. During the 1850s and 1860s, sporadic conflicts erupted between the Western Apaches and Anglo settlers, which led to the establishment of Fort Apache on White Mountain

THE CHIRICAHUA APACHES

Thomas E. Sheridan

On October 8, 1988, 102 years after Geronimo surrendered to General Nelson Miles, thirty-six Chiricahua Apaches from Fort Sill, Oklahoma, and the Mescalero Apache Reservation in New Mexico crossed the international border at Douglas, Arizona. Then they headed south into the Sierra Madre to reclaim their past. They were the grandchildren and great-grandchildren of people who rode with Juh, Loco, Nana, Naiche, and Geronimo. Many of them had never seen their homeland before.

For the next week they made their pilgrimage in sixteen vehicles, visiting places like Cañón de los Embudos, where Geronimo negotiated an earlier surrender with General George Crook, and Sierra Azul (Blue Mountain, "a sacred mountain," in the words of one of the pilgrims, Meredith Begay). They posed in several of the places where C. S. Fly, the famous Tombstone photographer, had photographed their ancestors more than a century before. "I never thought I'd be standing in the same footsteps as my great-grandfather," said Benedict Blake, the great-grandson of Naiche and the great-great-grandson of Cochise. "And there I was. It was a really good feeling" (Hayes and Hayes 1991:134).

From the late 1600s until 1886, the Chiricahua Apaches ranged across an enormous area stretching from the Mogollon Mountains in western New Mexico to the Sierra Madre of northern Mexico. They grouped themselves into three major bands—the Chihennes (Eastern Chiricahuas), who lived in New Mexico; the Chokonens (Central Chiricahuas), who inhabited southwestern New Mexico and southeastern Arizona; and the Nednhis (Southern Chiricahuas), who lived in Mexico. One smaller band, the Bedonkohes, occupied the headwaters of the Gila River. After the death of Mangas Coloradas in 1863, most affiliated themselves with the Chokonens and followed Cochise.

The Chiricahuas hunted deer, elk, and antelope; gathered acorns, yucca, mesquite pods, and agave; and cultivated small summer plots of corn, beans, and squash along mountain streams. But it was the raiding of livestock for food and mounts that brought them into two centuries of conflict with Spaniards, Mexicans, and Anglo Americans. Chiricahua chiefs like Mangas Coloradas and Cochise were master diplomats as well as war leaders, playing the state governments of Sonora and Chihuahua—and later the national governments of Mexico and the United States—against one another as they slipped back and forth across borders that had no meaning to the Apaches themselves.

With the coming of the transcontinental railroad—the Southern Pacific—in 1880 and the mining and cattle boom that followed, however, relentless pressure by Mexican and Anglo American troops guided by Apache scouts wore down most of the Chiricahuas. Beginning in 1876 the U.S. government confined Chiricahuas as well as Yavapais and Western Apaches on the hated San Carlos Reservation outside Chiricahua territory. Geronimo himself lived at San Carlos between 1877 and 1881. He returned for several months in 1884 after negotiating a surrender with General George Crook but fled to the Sierra Madre with Naiche and 132 other Chiricahuas in May of that year.

Crook mounted another campaign two years later, using Chiricahuas as scouts. But negotiations broke down after Crook's superiors insisted upon unconditional surrender and deportation to the East. Geronimo, Naiche, and forty other Chiricahuas slipped away, holding out in the Sierra Madre until late summer even though General Nelson Miles, Crook's successor, deployed 5,000 troops—one-fourth of the entire U.S. Army—to capture or kill the little band. After two Chiricahua scouts—Kayitah and Martine—tracked him down again, however, Geronimo decided to surrender for the final time on September 4, 1886.

Then the Chiricahua exile and diaspora began. In April 1886 the government deported seventy-seven Chiricahuas to Fort Marion on the Atlantic coast of Florida. Federal officials also sent thirteen Chiricahuas (including Chato, the Chiricahua war leader who became one of General Crook's most trusted scouts) to Indian Territory in Oklahoma to persuade them to resettle there. Chato refused. When he met with President Grover Cleveland in Washington, D.C., Chato demanded a "paper" allowing the Chiricahuas to remain in Arizona. Cleveland gave the illiterate scout a document welcoming him to the nation's capital. After the fanfare ended and the Chiricahuas were returning to Arizona, they were arrested as prisoners of war in Fort Leavenworth, Kansas, and shipped to Florida as well. Naiche, Geronimo, and their followers joined them later that year. By the end of 1886, 498 Chiricahuas were sweltering in dank, malarial Florida prisons, where 119 of them died.

Eight years later, in 1894, the government sent the Chiricahuas to a reservation at Fort Sill, Oklahoma, where they settled in "villages" that resembled their traditional local groups.

Government agents gave them animals to raise and seeds to plant, and most became ranchers and farmers. But in 1913, after twenty-seven years of exile, 187 of the 271 surviving Chiricahuas, including Chato, chose to leave their farms in Oklahoma and move to the Mescalero reservation in south central New Mexico. There, at least, they were close to the Mogollons, the Chiricahuas, the Dragoons, the Sierra Azul, and other mountain homelands.

Apache lands in 1868. A truce was quickly agreed upon with the White Mountain and Cibecue groups, and many men even served as army scouts in the bloody campaigns that followed against the Tonto groups, the Chiricahuas, and the Pais.

In 1874 the federal government began its removal program, marching thousands of Apaches from four reservations to the San Carlos subagency on the White Mountain Reservation. For a people whose subsistence depended on mobility, being forced to settle in a small area was devastating. In addition, the close association of more than twenty different bands that had not always been on the most friendly of terms led to tension and further unrest. The arrival of the Chiricahuas and Yavapais resulted in open hostility. It is not surprising that leaders such as Victorio and Geronimo left the reservation and led their bands back to their home territories or on raids into Mexico.

Geronimo and members of his Chiricahua band surrendered to the army in 1886 after a sixteen-month foray into southern Arizona and Mexico. Most Chiricahuas, even those like Chato who had served as army scouts, were deported to Florida, never to return to Arizona despite government promises to that effect. In the late 1880s the White Mountain, Cibecue, and Tonto groups were allowed to return to their own reservations. The Southern and Northern Tonto populations had been decimated, however, and ultimately combined on a tiny reservation. The Eastern White Mountain band was split, with some staying at the San Carlos Reservation and others returning to Fort Apache. The Cibecue group combined with the White Mountain Apaches to become the White Mountain Apache Tribe.

Today, the Western Apache people live on three reservations in east central Arizona. The northernmost, Fort Apache Reservation, is home to more than 10,000 enrolled members of the White Mountain Apache Tribe. The land, mostly above 5,000 feet in elevation, has vast tracts of mountainous forest lands that provide an economic base for logging and recreation enterprises. The adjoining San Carlos Reservation is slightly larger, with 1,826,541 acres but with fewer residents, 7,826 Apaches. Cattle ranching is the most important economic activity in the predominantly juniper and scrub oak vegetation zone. The Tonto Apache Reservation near Payson is one of the smallest of the reservations in Arizona, with eighty-five acres and ninety-two enrolled tribal members. Despite military defeat and internment on reservations, however, the Western Apaches have never lost their physical and spiritual connections to their mountain homelands.

Life in the Mountains

For centuries the mountains of eastern Arizona provided the Western Apaches with all the resources necessary for life. Their land was one of tremendous diversity. Remembering his childhood, Joseph Hoffman recalled that life was centered on the mountains: "In those days we had no flour and other food that we have now, but we had the fruit of different kinds of cacti, corn, and mescal. We used to eat the fruit of the prickly pear cactus and lots of other kinds of plants. It was God who made the earth and it was he who in the beginning told us to eat these different kinds of plants. This is the way that I learned how our people did" (Basso 1971:76).

The mountainous region covers five major vegetative zones, which include an abundance of plants and animals for use as food, medicine, construction materials, and tools. The rugged terrain also provides fertile valleys for farming and steep canyons that allowed the Indé to hide from their enemies. On another level, the mountains provided spiritual power to ensure success in healing, hunting, and warfare.

In the lowest elevations, the traditional lands of the San Carlos and White Mountain Apaches teem with the prickly pear and saguaro cacti of the Sonoran Desert. In the semidesert grasslands, from 3,500 to 4,500 feet, the perennial staple mescal (*Agave* spp.) grows among mesquite trees and a variety of grasses. Acorns and pinyon nuts can be gathered in the oak-pinyon-juniper woodland. Above 7,000 feet the pine forests offer a variety of large and small game.

Armed with an intimate knowledge of their environment, Western Apache families and bands moved their camps with the seasons to take advantage of the resources in the various regions. Grenville Goodwin, an ethnographer who interviewed many elderly Apaches in the 1930s, estimated that, in prereservation

A Western Apache (San Carlos) woman cutting mescal in about 1930.
(Smithsonian Institution, National Anthropological Archives)

days, 75 percent of their diet came from a seasonal pattern of
hunting and gathering.

After the winter snows had ended, bands moved to areas in
the semidesert grasslands where mescal plants were plentiful.
Families might spend several weeks gathering the plants' fleshy
hearts, their most important staple food. Women used a long oak
stick to sever the plant's roots and pry up the multileaved crown.
Then they cut off the leaves at the base, exposing the starchy
heart. While the women gathered the hearts, the men dug a
large roasting pit two to four feet deep and twice that in diameter. A large oak fire was laid, covered with stones, and after the
appropriate ceremony, it was ignited. Two or three dozen mescal
heads were placed on the coals, covered with wet grasses and
earth, and roasted for two days. When cooked, the hearts were
mashed and sun-dried on flat racks. Although the hearts were
sweetest and fullest in the springtime, they could be harvested at
any time and stored for more than a year. They were even a
valuable trade item, with a bundle of dried pulp commanding a
medium-sized Navajo blanket in return.

As spring and summer progressed, individual families moved
to their personal sources of prickly pear fruits, sunflower seeds,

An Apache family moving camp in about 1900. (Photograph by Edward S. Curtis, Arizona State Museum)

mesquite bean pods, and grass seeds. Women stored surplus food in baskets (see pl. 10), buckskin sacks, or pottery vessels placed in a cave or a *gowa*—an Apachean house made of bent, woven branches, covered with interwoven grass—until needed during the winter.

In the late spring, most Western Apache families would plant a small crop of corn, beans, and squash in the broader canyons, but this normally sedentary activity was just another part of the seasonal round. In 1932, Elder John Rope, a White Mountain Apache, described how farming had fit into their mobile life-style:

> When the corn was up about one and a half feet tall, it was watered once more. At this time it was beginning to form ears, and when it reached this stage, our people used to go off south of Black River to gather acorns, and the corn was left to mature by itself. . . . After a while they always sent someone back to Cedar Creek who would see how the corn was getting on and return to tell the others. If the corn was ripe, all our people would pack up the acorns that had been gathered and move back to harvest corn. (Basso 1971:95–96)

Late summer and autumn were busy times. Large groups of families traveled by foot and on horseback to the oak-pinyon-juniper woodlands, where women gathered the acorns of several species of oak that matured in August. This popular food was prepared in several ways: ground into a flour, shelled and eaten raw, boiled, or parched. A good acorn harvest was often the key to winter survival if the corn had not done well. In October and November, pinyon nuts and juniper berries were available at the same elevations.

Hunting was a year-round activity for men and boys, but in fall the meat and hides were at their best. Hunting parties ventured into the high forests to stalk deer and elk, and onto the high grasslands to pursue pronghorn. Stone- or metal-tipped arrows and mulberry bows were the preferred weapons. Lances were occasionally used to spear cows or hibernating bears. Young boys honed their hunting skills around the camp, using small bows and arrows to bring down rabbits, squirrels, and rats. By the time a teenager accompanied the men on his first deer hunt, he was usually an accomplished marksman:

> There used to be lots of deer in our country, and we ate the meat of them when we could get it. Near where we lived we used to hunt them. One man would go up on the top of a hill, and four or five other men would drive the deer up to him so that he could shoot them with arrows. About all that we did in

those days was to hunt and fight. (Joseph Hoffman, 1932; quoted in Basso 1971:79)

Around the time of the first snow, most families moved back down to the gowa of their winter homes in the semidesert grasslands. Here they could take advantage of the only year-round plant food: mescal. Mescal gathered in the mountains served many uses besides food. The ground, roasted heads were boiled and the liquid fermented to make a mildly intoxicating drink similar to *tiswin,* a beverage made from corn. Apache fiddles (pl. 11) were fashioned from a section of the pithy flower stalk that they hollowed out and fitted with one or more horsehair strings. Their music was commonly heard at dances and ceremonies. The bulbous base of the flower stalk could also be hollowed out for use as a water container. For sewing, the spine at the end of each leaf was used as a needle, with the long leaf fibers left attached to serve as a thread.

Perhaps the most versatile plant used by the Western Apaches was beargrass (*Nolina* spp.). Its sturdy but pliable leaves were used to make everything from houses to cradleboards. For use as a covering for the gowa, whole beargrass plants were tied to the wooden framework in overlapping layers, much like a thatched roof in other parts of the world. Although the gowa was often wrapped in canvas when available, the beargrass covering was adequate in moderate rains. Women plaited the leaves for use as temporary cradleboard hoods, food strainers, and mescal trays. The scooplike base of the leaf made a handy spoon when it was dried and hardened. For washing hair, Apache women mixed the crushed heart of the plant with water to make an effective soap.

The division of labor in an Indé family was well defined. The gathering of plants was strictly the province of women. As Joseph Hoffman pointed out, men took on few subsistence chores aside from hunting. They provided the heavy labor of digging mescal roasting pits, but only the women dug up and trimmed the hearts. For the most part, women prepared the food, tanned hides, made most of the subsistence tools, baskets, and clothing (see pls. 12–14), cared for the children, and did the heavy work that kept the family home operating. Men and women shared the work involved in planting and harvesting the summer gardens.

A Western Apache cornfield in 1936. (Photograph by Grenville Goodwin, Arizona State Museum)

In their seasonal rounds, Western Apaches covered vast ranges of territory. The mountains provided a variety of resources that were simply unavailable in less diverse terrain. Through their intimate knowledge of the flora and fauna, matched with a mobile lifestyle, the Indé met their physical needs in an efficient and effective manner.

Power from the Mountains

The mountains not only provided the sustenance for Apache life in the physical sense but were also the source of many spiritual objects that brought power to the Indé. The supernatural power that is present in the universe is the overriding spiritual concept of traditional Western Apache belief, and it informs most of the Apaches' religious ceremonies.

While the mountains themselves are not imbued with spiritual power, Apaches believe that many objects associated with them possess an invisible and potent force that can act on humans in many different ways. Animals such as the deer, bear, and mountain lion; quartz crystals, turquoise, and prehistoric artifacts found in mountain canyons; and the lightning associated with summer thunderstorms rolling in from the cloud-covered peaks—all of these have power that can be controlled by individuals and used for either good or evil.

The Western Apache concept of *diyiʾ* describes a type of force that is a part of certain animals, plants, stars, and natural phenomena, such as wind, rain, and lightning. This force is as much a part of the object as its physical body. The power of these objects resides largely in a supraterrestrial region of the world, but it can be used by people who are fortunate or persistent enough to tap a small part of one of these universal forces for their own use. As one unidentified Apache man noted,

> There is a kind of power used in wrestling, in which a man having it can throw a man twice his size if he uses the power on him. He does not do it by his strength at all, but by his power.
>
> I have heard that men knowing this when they were going to wrestle, as they approached their opponent, would say, "On

A woman harvesting corn at East Fork in about 1900. In the words of Joseph Hoffman, "When the corn became ripe at that place where our people had planted, we harvested it, and then all went on to Cibecue to where there were lots of farms." (Photograph by Arthur Guenther)

my right side is mountain lion holding up my arm, and on my left side is another mountain lion holding up my arm." (Quoted in Basso 1971:273)

In some cases, a power seeks out a person, presenting itself through repeated lightning strikes or a potent dream. Keith Basso recorded the following description of the accentuated dream in which a Cibecue Apache man gained "bear power":

Long ago it happened this way. I lay down in my wickiup and saw a big mountain. Only I didn't know I was sleeping. It was all covered with spruce, that mountain, and I thought I really saw it. Then, on the top I saw something, so I went up, and it was a bear. I was sure scared. But then that bear talked to me. At first I didn't know what it was saying, but then I started to understand. It wasn't the words we people use around here. So it talked some more and I heard it. After that I wasn't scared. I followed it to a spring and we drank some water. Then it talked some more. I didn't say anything. Pretty soon that bear told me that if I went to a certain place I would find a black stone. That bear told me to get it. After that, the bear went away. And I tried to follow it but it didn't leave any tracks. Then I woke up. I was sure scared when I knew what I had been dreaming. I wondered about it. Then, in the morning, I went over to the place where that dream I had told me to get the stone. I found it right away, and then I knew that I could get some of that "Bear Power." (Basso 1970:41)

Another Cibecue individual described the manner in which he was chosen as a recipient of "lightning power"—interpreting an unusual event as a symbolic and highly charged experience:

Over there at Gleason Flat, one time, I was working cattle. That day was sure hot. Then, just when I was looking in that brush country for some cow, it started to rain. Sure rain hard. So I went with my horse under a tree. Right then some lightning hit about from here to that house . . . and start rolling at me. I was sure scared. But then that lightning stopped and went into the ground just when it got to me. Then, another lightning did just like that. Come at me and go into the ground. After that I thought about it. When I got back to Cibecue, I started learning songs. I knew that power ("Lightning Power") wanted me to use it. (Quoted in Basso 1970:41)

In other cases, an individual may decide to seek a certain type of power and arrange to serve an apprenticeship with a person who has already learned to control the power to heal sickness or prevent misfortune. After making a connection with a diyiʔ, the recipient has to learn the extensive and complex body of chants and prayers that have been a part of that power since the earliest time of creation. These incantations are nearly inseparable from the power itself—both are referred to by the term *diyiʔ*. The practitioner must exercise considerable skill in "singing" these chants in order to control and channel their power effectively. Western Apache curing ceremonies, certainly the most frequently performed ceremonies until recent times, usually consist of a long recitation of dozens of chants, a process lasting from sundown until dawn. The chants are crucial. As one of Basso's consultants explained, when a power hears its songs it will be compelled to listen. If a person does not sing the appropriate songs, a power will not know where to find him, and it will not want to work for him.

Some sources of supernatural power are typically associated with specific abilities and desired results. For instance, Bat Power conveys the ability to dodge bullets in battle. But in most cases when a person is chosen for a power, she or he must experiment to find out its specific uses and effects. Through trial and error, one may discover an ability to cure certain illnesses, to find lost objects, to run great distances, or to be successful in hunting. Often the power allows a person to diagnose and cure illnesses that are due to unfortunate accidental contact with other powerful spiritual entities, such as by drinking water in which a bear has been swimming or breathing smoke from burning wood on which a bear has urinated. Most commonly, the power has the ability to ward off misfortune of all types, bringing success, health, and long life to its possessor.

Probably the most common diyi⁊ in the mountainous homelands is lightning. The sudden and frightening appearance of a lightning strike, often with devastating results, is symbolic of the way in which an accidental contact with uncontrolled and potentially negative power manifests itself. While a very few individuals may learn to control this power, many fear the illness and death that may result from direct or indirect contact. A close strike is reason enough to schedule an expensive curing ceremony to prevent potential illness. Lightning power is so dangerous and pervasive that one can even get sick from merely breathing the smoke from firewood that had been struck, or eating beef from a steer that had a close brush with lightning while still alive.

The power of lightning can also be transferred to other objects found in the mountains. The San Carlos and Fort Apache Reservations contain many prehistoric sites of the Mogollon people who lived in the region from about A.D. 100 until about 1400, before the Western Apaches arrived. It is not uncommon to find ancient arrowheads, stone knives, arrow straighteners, or ceramic pipe bowls among the ruins. While Indé recognize that these objects belonged to the prehistoric residents of their mountain homeland, the stone points are also said to be the result of lightning strikes in the forest. A war charm necklace strung for anthropologist Grenville Goodwin in 1935 contains forty-one prehistoric projectile points and flakes left when stone tools were made. Such necklaces were commonly made and worn in the Southwest through the 1880s (Ferg 1987). It is almost certain that Western Apache soldiers in more recent wars have carried similar charms. In 1942, seven soldiers from the village of Cibecue were blessed with Bat Power before leaving for the front. All returned home alive.

Quartz crystals from the mountains offered a special opportunity for those who could master their power. Like other charms, they could ward off illness and misfortune. A practitioner gazing into the translucent core could also see into the future to foretell events or locate misplaced objects.

While the concept of power and the knowledge of the chants, prayers, and taboos associated with it may not be as prevalent today as fifty years ago, it is still an important part of Indé belief and culture. Lightning is a force to be reckoned with, and its potential for illness is not ignored. Western Apaches today readily accept treatment at modern health clinics but may also turn to a curer to diagnose and treat the deeper spiritual cause of an illness. It is not uncommon for diseases such as cancer to respond to curing ceremonies as well as chemotherapy. The power of Apache beliefs and traditions still runs strong in our modern world.

Stalking with Stories: The Moral Power of the Mountains

In 1979, Benson Lewis, a White Mountain Apache, told Basso, "I think of that mountain called 'white rocks lie above in a compact cluster' as if it were my maternal grandmother. I recall stories of how it once was at that mountain. . . . Stories make you live right" (Basso 1987:96–97). A year later Nick Thompson said, "The land looks after us. . . . So we stay away from badness" (Basso 1987:95–96). The statements by Benson Lewis and Nick Thompson, respected Western Apache elders, are intriguing and somewhat confusing to non-Indian people. It is difficult for most citizens of the United States to comprehend why someone would equate an outcropping of rocks on a nearby mountain with a close relative, or how the mountains could make someone live right. But the Indé have a relationship with the landscape that is very different from that of most Americans of other cultural backgrounds. The mountains are not only a source of supernatural power, they are also a source of moral power that still holds sway over people who know the land and the stories it can tell.

The key to understanding these statements is the existence of a body of traditional Apachean stories called 'ágodzaahî nagoldi'— "to tell of that which has happened." These 'ágodzaahî nagoldi' constitute a major body of Western Apache oral literature along with stories of the creation (godiyįhgo nagoldi', "to tell the holiness"), sagas (nlt'éégo nagoldi', "to tell of pleasantness"), and stories that arise from gossip (ch'idii). Keith Basso describes the 'ágodzaahî nagoldi' as historical tales whose purpose is to criticize, to warn, or "to shoot." Nick Thompson says of these stories, "We shoot each other with them, like arrows" (Basso 1987:102). Each story conveys a moral lesson, and elders frequently must shoot an errant young relative with an 'ágodzaahî nagoldi' to convey important values of the Apache culture and "keep them living right." But how do these stories become equated with the mountains, and how can the mountains "become" that elder relative?

Each story relates an event that happened at a certain place on one of the Western Apache reservations. Indeed, the tales follow a specific format: each begins and ends with the highly descriptive Indé name for that place. Young people growing up in Cibecue or Whiteriver first learn the names of dozens of significant landscape features in the surrounding region. As they mature, they learn the events that are associated with these places and ultimately the lessons that can be learned from them. For example, the following story, told by Benson Lewis, recounts an event that happened about a hundred years ago near Cibecue. It carries a message that is still very relevant today: one should not act so much like a white man that one turns against the Apache people.

It happened at "men stand above here and there." Long ago, a man killed a cow off the reservation. The cow belonged to a

Whiteman. The man was arrested by a policeman living at Cibecue at "men stand above here and there." The policeman was an Apache. The policeman took the man to the head Army officer at Fort Apache. There, at Fort Apache, the head Army officer questioned him. "What do you want?" he said. The policeman said, "I need cartridges and food." The policeman said nothing about the man who had killed the Whiteman's cow. That night some people spoke to the policeman. "It is best to report on him," they said to him. The next day the policeman returned to the head Army officer. "Now what do you want?" he said. The policeman said, "Yesterday I was going to say HELLO and GOOD-BYE but I forgot to do it." Again he said nothing about the man he arrested. Someone was working with words on his mind. The policeman returned with the man to Cibecue. He released him at "men stand above here and there."

It happened at "men stand above here and there." (Quoted in Basso 1987:107)

The details of this story can be understood only in the context of the restrictive life of the Fort Apache Reservation, which was imposed on the Apache people in the late nineteenth century. They were confined to areas around the army posts, and the ability of families to hunt and gather foods was severely limited and hunger was ever present. From an Indé perspective, killing a cow belonging to a white rancher could hardly be considered a crime. The Western Apache policeman should have known this and should never have taken the man to the army officer. The policeman's confused words were the result of a spell cast by someone with power in order to foil his loyalty to the officer.

Lasting bonds are created between individuals and places, as illustrated by a story Keith Basso first heard in June 1977. It was told by a grandmother to a young woman who had worn large pink curlers in her hair when she went to a puberty ceremony at Cibecue. Although the curlers were considered stylish at her BIA boarding school in Utah, the seventeen-year-old was old enough to know that Apache women wear their hair loose at such events to show respect for the ceremony. Everyone took notice of her actions, but nothing was said at the time. Two weeks later Basso attended a birthday party at the girl's grandmother's camp. After dinner, as the relatives sat around talking, the grandmother told the story of the policeman who acted like a white man. A few minutes later the young woman stood up without a word and left. When Basso asked her grandmother what had happened, she replied, "I shot her with an arrow" (Basso 1987:110).

Nearly two years later, Basso happened to give the young woman a ride home from the trading post at Cibecue and asked her about the story. The event was still a vivid memory. She said that at first she thought that maybe her grandmother was telling the story for someone else. Then she thought about the ceremony and knew it was for her. "I sure don't like how she's talking about me, so I quit looking like that," she said. "I threw those curlers away" (Basso 1987:110). As they passed the place called "men stand above here and there," Basso pointed it out. The young woman smiled and said quietly in Indé, "I know that place. It stalks me every day" (Basso 1987:110). For the rest of her life, she will not see that spot in the mountains without thinking of her grandmother, the story, and what it means to be Apache.

These ageless mountains and the stories they tell can still contain powerful messages that shape new generations of Apache youth. But in the words of White Mountain Tribal Chairman Ronnie Lupe, "Our children are losing the land. It doesn't go to work on them anymore. They don't know the stories about what happened at these places. That's why some get into trouble" (Basso 1987:95).

A New Relationship with the Land

Nick Thompson and Ronnie Lupe are undoubtedly correct in their observation that many Western Apaches are not as close to the land as they were fifty years ago. They no longer spend most of their time combing the canyons and hillsides in search of food and other resources, and they have lost much of the intimate body of knowledge of the plants and animals in the mountains that had been handed down for generations. Yet even in the towns of Whiteriver and San Carlos, communities with modern medical clinics, schools, new houses, video rental stores, and restaurants, the mountainous lands remain the basis of the

Western Apache way of life. As with any culture, changing technology and conditions dictate new adaptations, but important values persist.

For centuries, the mountains have provided the Western Apache people with more varied natural resources than most other Indian tribes in Arizona. Through wise management of these resources, the Indé have achieved an economic base that maintains their ties to the land and preserves their enduring cultural values.

For generations of San Carlos Apaches, living close to the land has meant raising cattle (Getty 1963). The San Carlos Reservation produces more "cowboys" than any area of comparable size in Texas or Oklahoma. San Carlos cattle associations run cattle on the open range through the rugged canyons and grassy hills of the Natanes Plateau. Apache cowboys head out each spring and fall on roundups, searching every arroyo for calves that will soon feel the sting of a hot branding iron.

Today, cattle operations at San Carlos are a unique blend of old and new, Apache and Anglo. The beginnings of Western Apache ranching are rooted in the ration system on the reservations in the 1870s, when some people accumulated their weekly allotments of beef so that they could eventually obtain a live steer. By the 1880s the Indian agency was issuing cattle with the intention of starting a viable industry among the San Carlos. Today, more than 20,000 head of cattle graze the 1.6-million-acre reservation (Buskirk 1986).

The bulk of this livestock is worked through five cattle associations, cooperative organizations in which cattle are owned individually but worked communally by the group's members. A cow will carry the brand of its owner as well as that of the association. The cooperatives each have a geographic base, but membership is also determined by clan and family ties and other social relationships. A board of directors is elected to manage the operations, and a certain amount of prestige accompanies these positions. Formalized in the 1930s, the association system maintains some aspects of the consensus-oriented, band-based leadership of prereservation life.

At the same time, the San Carlos Apache Indian Community has brought in the highest level of expertise available to improve the breeding stock of reservation cattle. Since 1956 the tribe has worked with personnel from the University of Arizona's animal science department and the u.s. Department of Agriculture to develop a herd of quality purebred and registered Hereford cattle, using the latest techniques in animal husbandry. The tribal herd serves as a cooperative experiment station, with scientists conducting thousands of laboratory tests each year to evaluate the offspring of different dams and sires. The project has resulted in one of the best herds of range Herefords in the world, and its superior breeding stock is in demand not just among the San Carlos associations but on ranches throughout the Southwest.

The White Mountain Apaches have found other ways to manage their mountain resources and provide employment for their people as well. In so doing, they have had to assert their independence and authority to control their land and their destiny.

Arizonans can hardly think of outdoor recreation without conjuring up images of the White Mountains: crystal-clear mountain streams, alpine lakes brimming with trout, meadows grazed by elk, forests stalked by bears, and high mountain slopes cleared for skiing. Since the 1950s, the White Mountain Apache Tribe has implemented a careful plan to protect and develop their resources, allowing access to outsiders for the benefit of tribal members.

In 1957, tribal council members at Fort Apache risked arrest by ordering the construction of a dam on a reservation stream to create Hawley Lake at an elevation of more than 8,000 feet. Anglo farmers downstream around Phoenix filed suit to stop what they saw as a threat to their water. The Apaches persisted and ultimately prevailed. Today, twenty-six lakes, 420 miles of trout streams, and more than 1,000 campsites lure hunters, anglers, campers, and backpackers to the Fort Apache Reservation, generating hundreds of thousands of dollars in revenues and dozens of jobs.

White Mountain Recreation Enterprises manages these resources so that the natural quality of the land is maintained as much as possible. While reservation guests are encouraged to enjoy themselves fully, activities such as fishing, backpacking, snowmobiling, sledding, and off-road vehicle use are regulated by permit and restricted to approved areas. Most campers are pleased to learn that only the quiet hum of electric outboard motors is allowed on Western Apache waters so that the serenity of a forest-lined alpine lake is not broken by the abrasive whine of a gas-powered motor.

In 1974 the White Mountain Game and Fish Department took control from the State of Arizona of the issuance of hunting permits on Western Apache lands. The number of elk permits was reduced from 450 per year to 32. The fee was increased from $35 to $3,000. As a result, the elk herd has grown to more than three thousand animals, while trophy hunters from across the country have enjoyed success rates of 95 to 100 percent.

The crown jewel of this Apache recreational enterprise is the Sunrise Ski Resort. Opened in 1970, the lodge and runs have become the premier downhill skiing area in Arizona, attracting visitors from many adjacent states to the slopes of Sunrise Peak. The enterprise has expanded steadily over the past two decades to include two lodges, a hotel, three restaurants, and an up-to-date double chairlift. More than 300 Apache employees now spend their winters in the deep, white powder of the high mountains. Many Indé have become expert skiers, taking advantage of the free skiing privileges extended to all tribal members.

Probably the most profitable use of mountain resources is the lumber operation conducted by the Fort Apache Timber Company (FATCO). Replacing outside operators who cut trees on reservation lands, the tribal corporation was formed in 1963 to develop a lasting economic base for the White Mountain Apache Tribe. Timber lands are managed for the long term, with approximately 80 million board feet harvested annually on a sustained-yield basis that emphasizes the thinning of forests through the harvesting of older mature trees. In addition to logging operations, FATCO operates three lumber mills in Whiteriver and one in Cibecue. In 1990 the company provided 307 jobs in the forests, mills, and greenhouses, a significant economic impact for the 10,394 tribal members. But the benefits do not stop with new jobs. FATCO profits have funded the Mary Riley Education Building, the new Whiteriver Shopping Center, and other projects that benefit all of the White Mountain Apache people.

As they have done for centuries, the mountains still provide spiritual renewal and physical sustenance for the Indé. Whether operating a ski lift, driving a logging truck along a mountain road, or drinking from a clear stream on a hot summer day, the Indé feel at home in the mountains. Recent decades have brought many changes for the Apache people, but some things never change. As Nick Thompson noted in 1980, "We used to survive only off the land. Now it's no longer that way. Now we live only with money, so we need jobs. But the land still looks after us" (Basso 1987:95–96).

THE HAVASUPAIS, HUALAPAIS, AND YAVAPAIS
THE GREAT CREATOR HAS GIVEN US THIS COUNTRY

Nancy J. Parezo

For centuries, the Upland Yumans—Pais (Pa'a) and Yavapais—occupied nearly a third of northern and central Arizona, from the Colorado River to the San Francisco Peaks and from Chino Creek to the Pinal Mountains. These peoples have deep attachments to this area, which is part of their cultural identity. As then Havasupai vice-chairman (now chairman) Rex Tilousi said in 1988, "I am related to this land. We believe in our Mother Earth, who brings forth the water, plants and animals we need; our Father Sun and Uncle Moon. I have ties with all those. I feel I am related to them" (*Arizona Daily Star*, May 29, 1988).

But others also wanted the beautiful land. Since Anglo Americans first threatened Yuman territories in the mid-1800s, the Havasupais, Hualapais, and Yavapais have fought a continuous war to retain and protect their lands, heritage, rights, and way of life. More than a century later, their struggles continue in courtrooms and in the hearing rooms of Congress. This is a difficult undertaking, for native peoples find Anglo American attitudes toward the land difficult to comprehend and accept because of deep-rooted philosophical differences. According to John Williams, a Yavapai, "I don't know about the white people. I don't know who they are, where they come from. But we people don't come from nowhere across the ocean. We were raised right here in this country. . . . This is our home" (1975, quoted in Hinton and Watahomigie 1984:1).

It is this continuing sense of place that has helped the three societies survive in a difficult world. They remember that the Creator said in the beginning, "This land will be your country to settle and get your living in. . . . Mark off your land and live by the water" (Kuni quoted in Kroeber 1935:16–24). Pai land is holy land, a situation as real to the Pais as God's granting of Israel is to the Jews. The Hualapais, Havasupais, and Yavapais claim ownership of their land not merely in terms of occupancy but in the sense of their irrefutable right to it as chosen peoples.

The Hualapai Creation Myth

In the beginning there was a great flood across the land. Only Wikahme, Spirit Mountain (Newberry Mountain near Davis Dam on the Colorado River in southern Nevada), towered above the waters. The Giver of Life and Keeper of the Universe created the twins called Older Elder Brother, Matavila, and Younger Elder Brother, Judaba:h, and placed them on Spirit Mountain. They gathered canes ('ta), cut them into pieces, and placed them on the ground with the tops toward the east. When the brothers fell asleep, Matavila transformed the canes into people. The longest strip he called Mohave, and the next longest he called Hualapai. The twins gathered the new people together and instructed them in many things. Then they migrated south from Spirit Mountain. The Elder Brothers led them to the sacred canyon *watwita*. Here they planted corn, beans, and squash, and gathered agave, yucca, and pinyon.

> Judaba:h then told the people:
> "No far away lands,
> You are not to go or be anywhere;
> Here, the water that lies here
> The land here, the land along this river,
> Here, you roam, here,
> Be around here,
> You are to be here, it is destined,
> This is what you are to do, it is so," he said.
> Judaba:h did this, gave this to us, it is told.
> from Wikahme', told by Paul Talieje in
> Hinton and Watahomigie 1984:37–38

Judaba:h also told each band to find their own place: "Here is the land where you will live. Go to the places where you find water.

Mark off your land and live by the water. Name these places"
(Kuni quoted in Kroeber 1935:15–16).

Upland Yumans: Three Similar Cultures

Anthropologists refer collectively to the peoples who live in
north central and northwestern Arizona as the Upland Yumans.
All speak closely related, mutually understood languages. The
Spaniards considered all the peoples living in this area to be one
group, whom they called Cosinas, a Hispanicization of the Hopi
word *Ko'ho'-nin*. The term *Hualapai* comes from the Yuman
word *Xawalupa'ya*, meaning Pine Tree People. The Havasupais
refer to themselves as Ha vasuwa'baaje, which translates as People of the Blue-Green Water, and the Yavapais call themselves
Yavepe or Nyavpe', which means People of the Sun.

Spanish contacts with the Upland Yumans were minimal, and
their influence was small. Antonio Espejo, on his search for mineral wealth in 1582, met some Yavapais near Jerome. Since the
people wore small crosses dangling over their foreheads, he called
them Cruzados. Over the next century, explorers met people
whom they categorized as wanderers and nomads and called
them by many names, for example, Serranos (Yavapai mountain
dwellers), and Jumanas (Havasupais). The groups continued to be
confused, and after the 1690s the Yavapais were increasingly
referred to as Yavapai-Apaches or Mohave-Apaches. It was not
until 1776 that Francisco Garcés met thirty-four families in Havasupai Canyon and specifically distinguished three Upland Yuman groups—Walapais, Yabipais, and Coninas.

There were good reasons for Spanish confusion. In the past,
Upland Yuman groups each considered themselves one people,
"the only true human beings on earth," who later separated and
became autonomous. Hualapais and Havasupais continue to
consider themselves one people (Pais or Pa'a) who until recently
were organized into self-sufficient bands and subtribes that
could be mobilized in times of war. Until the Hualapais were
confined to reservations, their society consisted of three distinct
subgroups: Witoov Mi'uka Pa'a (Middle Mountain People), Yavapai Fighters, and Ko'audva Kopaya (Plateau People), who included the Havasupais. All subgroups intermarried and viewed

the Yavapai groups—Yavepe (central Yavapais), Tolkapaya (western Yavapais), Wipukpaya (northeastern Yavapais), and Kewevkapaya (southeastern Yavapais)—as traditional enemies.

No other groups in the Southwest held a larger and more diversified territory at historical contact than the Upland Yumans. Bands traveled from sun-baked deserts through sage and juniper forests to the pine forested slopes of mountains. Each culture systematically used all parts of these environmental zones, but each group specialized in a specific zone. The Yavapais lived in desert and rugged mountain areas, and the Hualapais lived in riverine areas and the rugged side canyons of the Coconino Plateau. The Havasupais lived part of the year in temporary villages located in the canyons of the Colorado River 5,300 feet below the rim. Here they grew corn, beans, sunflowers, melons, tobacco, and squash by floodwater irrigation and dry farming (which is farming that depends solely on groundwater and rainfall) on small plots in very difficult terrain. Having a permanent stream and fertile land at the bottom of steep, red rock canyons, the Havasupais grew more of their food than other Pais, but this was a distinction only of degree and scale, for all groups practiced some horticulture. The Yavapais farmed the least because they lacked favorable localities and often had to move when faced with retaliatory raids by enemies.

The Upland Yumans' large territories required great localized knowledge. Band distinctions were largely geographical, and each group had a recognized use area. A band permanently inhabited this area rather than particular local spots, a concept that Anglos never understood or respected. A settlement area included several localities that smaller groups of three to four related families visited annually on a distinctive seasonal round. Their rancherías were located near concentrations of resources and water, and family-owned farming plots and gathering sites were recognized. Some Pais and Yavapais continued these patterns in an abbreviated fashion into the 1940s, and most bands still have areas that are considered special.

Upland Yuman cultures were very similar to that of the Southern Paiute in subsistence patterns and technology. As with any people who have to move often, the material culture of each group was portable. Until the turn of the century, Hualapais and

A Yavapai family at their camp in about 1887. (Smithsonian Institution, National Anthropological Archives)

Havasupais made small amounts of pottery that basically re-sembled Hopi and River Yuman undecorated brownwares. Yava-pais made a distinctive ware called Tizon Brownware. Tools, knives, clay pipes, awls, fire drills, and hearths were also kept to a minimum. Heavy grinding stones and fragile pots were left at rancherías rather than carried. All groups made excellent bas-ketry—coiled trays or bowls, twined burden baskets, parching trays, cradleboards, water jugs, and cooking baskets (see pls. 17–19). Many household baskets were plain; on others, simple geo-metric and banded designs were worked in martynia (devil's claw), mulberry, yucca, and soapweed. The baskets of each group can be distinguished by design, small differences in construction techniques, and occasionally materials.

Upland Yumans made little investment in housing. Houses (*u-wa'*) were temporary mud-covered and willow-thatched structures built on a dome-shaped framework of poles, which early explorers described as similar to Western Apache wickiups. There were, however, several differences. As Odopaka described Hualapai houses to anthropologist Alfred Kroeber, "Houses

HOW TO WEAVE A BASKET

Nancy J. Parezo

Yavapai baskets (pl. 19) are very similar to Western Apache baskets, and Yavapai weavers have greatly influenced Apache designs and techniques. Designs are said to have been given to each weaver by the Great Spirit and include geometrics, diamonds, stars, and armed crosses, which the Yavapais call the "wheel of life." In the early 1970s, Grace Jimulla Mitchell, a Yavapai artist and politician, described how one should weave a basket. Mitchell was the daughter of Viola Jimulla, who was the chief of the Prescott Yavapais in the 1940s and 1950s. Mitchell followed in her mother's footsteps in 1967 and led the Yavapais until her death in 1976. This text was presented in *Spirit Mountain: An Anthology of Yuman Story and Song*:

A Yavapai woman making a basket in about 1900. (Smithsonian Institution, National Museum of Natural History)

I'm going to weave a basket. I'll gather mulberry shoots, split them and roll them, and then tie them together. Then I'll dry them. When they're dried, I'll boil them. I'll boil them until they're ready. Then I take them out. I'll peel them. I peel them and dry them, too. When they're dried, I'll split them, too. Then I'll scrape it. I'll arrange them to size so that they are ready for weaving.

I'll gather devil's claw, too. I'll get devil's claws which are long, black, and good. When they're ready in small bunches, I'll tie them in small bundles. Then I'll weave a basket.

I'll gather long shoots, gather cottonwood and willow. I'll gather mulberry, too. I'll bring them home. Then I'll scrape them. Then I'll tie them in bunches. After doing this, I'll weave a basket.

I'll get devil's claws. With them I'll weave at the base. They will come up from the base. Devil's claws strengthen and hold it fast as it spreads out. On top I'll weave with white ones, too. I stick the long shoots in it. Then I weave.

Then I put in designs. When I have finished this, I weave the edge with the devil's claws as the last thing. When I have reached the point where I began, I will have finished. Devil's claw is strong; it will hold fast. (Hinton and Watahomigie 1984:176)

Today the desire for cultural preservation, the Yavapais' historical trading patterns, and the demand for American Indian art have kept traditional basketry alive. Some of the Upland Yumans' biggest customers are the Hopis, who often line the interior walls of their houses with Yavapai, Havasupai, and Hualapai baskets.

faced east and were made by men" (1935:125). A hearth was placed in the center, and "the family slept opposite the door and the guests slept along the sides of the house." The Havasupais also made a rectangular house with a brush top, and during the late nineteenth century they adopted a modified version of the Navajo hogan and sweat lodge. Like the Yavapais, they utilized caves for shelter in the winter because they held heat well and made ramadas for shade in the summer. Throughout the twentieth century, many Upland Yumans have lived in Western-style housing, although some lived in traditional houses into the 1930s and 1940s.

Pai political organization was not highly structured, although there were leaders (*pa-kawha't*) in each band who served as spokesmen and influenced others by their wisdom, diplomatic and judicial skills, experience, and knowledge. As Blind Tom told Kroeber, "A pa-kawha't is a man who speaks or talks, hence 'chief'. It was a higher position and so commanded a certain amount of respect from ordinary men. . . . The office went in the male line, but not if a chief's son was a weak man" (1935:153). Both men and women informally voted on the candidate. Although Hualapai leaders oversaw band and camp activities, they respected the right of individuals and families to make their own decisions. Even in time of war, the most competent individual among the headmen led what was essentially a defensive or raiding unit composed of volunteers. Nevertheless, the Hualapais could and did quickly mobilize up to a quarter of the tribe, a remarkable feat given the distances involved.

Some Upland Yuman cultural differences stemmed from contacts with other American Indian groups. All were great traders. Historical trade routes had prehistoric counterparts, which indicates long periods of interaction. For example, Cherum, a Hualapai headman in the 1850s and 1860s, traded with the Hopis and Zunis for woven textiles, which he in turn traded to the Mohaves for horses. He then exchanged horses for firearms with the Southern Paiutes, who had obtained the weapons from Mormons in Utah (Dobyns and Euler 1970:23). The Yavapais traded dried mescal and meat for corn with the Pimas, and baskets, buckskin, and mescal to the Tohono O'odham. Some Yavapais also traded buckskin, baskets and ram's horns with the Hopis.

A Havasupai settlement, undated. During the hot summer days, nearly all activities took place outdoors or under ramadas. The thatched houses were used only for sleeping. (Photograph by Jesse Bratley, San Diego Museum of Man)

Yavapais and Hualapais also participated in rarer expeditions to the West Coast, south into Mexico, north to Paiute country, and east to Zuni. All groups traded with the Navajos in order to obtain their famous blankets. Both the Havasupais and the Hualapais acquired Pacific seashells from the River Yumans and mined red hematite (for paint) to trade with the Halchidhomas on the Colorado River, who were later forced to move up the Gila River by their enemies, the Mohaves and the Quechan.

Some bands were greatly influenced by and simultaneously enriched other Indian societies through their mutual contacts and occasional intermarriage and military alliances (to stem Spanish slave raiding). The Havasupais, for example, interacted extensively with the Hopis. The Hopis' annual salt expedition to the Grand Canyon brought them to Havasupai land, and it was not unusual for Hopi families to live with the Havasupais during times of drought or for Havasupai individuals and families to live with the Hopis of Third Mesa during the winter months. The Hopis were also military allies. In 1583 the Hopis asked the Havasupais to help fight the Spaniards. Hopi elders decided not to attack the Espejo expedition, however, and the Havasupais returned to their canyon. As a result of these contacts, the Havasupais used many Hopi items—pottery, cotton, tobacco, blankets, jewelry—and the Hopi obtained baskets, foodstuffs, meat, and skins in return. Anthropologist Alfred Whiting learned that in the 1870s the Havasupais gave lima beans to the Hopis and that the Hopis reciprocated with peach trees and melons. The Havasupais modified Hopi Kachina dances into individually sponsored affairs conducted to bring good fortune and prosperity to the people. The largest ones were performed in late summer or early autumn as a type of general reunion held before the people ascended the plateau for the winter.

The Struggle to Survive

According to historian Elliot Coues (1900), the main effects on the Hualapais of Spanish colonization in Arizona were a slight shifting of residence east, an increase in trade and social interaction, and the importation of European goods such as Spanish belts, awls, cloth, and metal tools. The Hualapais obtained these items from the Hopis, however, not through direct trade. In general, the Spaniards (and later the Mexicans) and Upland Yumans had little contact, but most of the contact they did have was friendly, with the exception of that with Spanish slave raiders.

The same cannot be said of the interactions between the Upland Yumans and Anglo Americans. The first contacts with French Canadian and Anglo American trappers were recorded in 1826. Thereafter the Upland Yuman bands tried to avoid them, but this became increasingly difficult with the gradual but persistent curtailment of each society's territory. On the east, Navajos began to occupy the lands between Hopi towns and the easternmost Havasupai camps. By 1842 the Pais had begun to engage in intermittent retaliatory attacks to stop incursions onto their land. In 1851 the Cerbat Mountain Pais, for example, attacked Captain Lorenzo Sitgreaves's guide, Antoine Leroux, west of Flagstaff. Two years later, however, the Sitgreaves party passed safely through the same area, as did Lieutenant Edward Beale, who stayed briefly at Peach Springs in 1857–58.

It was not until the 1860s that continuous contact resulted in trauma and the rapid loss of Upland Yuman lands. U.S. immigrants began to travel through Hualapai lands in 1858, and the Pais collected a toll in livestock in return for safe passage. When gold was discovered in Prescott in 1863, prospectors quickly scattered throughout Hualapai territory. Ranchers followed in their path, and their herds of cattle destroyed wild plants and game resources. "All at once just out from nowhere there were white ranchers, homesteaders coming to this part of the country," Mary Wescogame told the federal Indian Claims Commission. "The whites told [us] to get out, never to come back" (1950:72). When the Pais killed mules, horses, and oxen for food, settlers demanded that the army relocate them to military reserves.

Several Yavapai bands were captured and moved to Camp Verde in 1861. Soon after this, approximately 2,000 Hualapais were forced to settle on the Colorado River Reservation, but they soon found that they had to leave or starve. They returned to the mountains but met with local Anglo hostility in 1865. In April, secure in the knowledge that a local federal district judge believed that "a white man should never be punished for killing an Indian," settlers killed a band headman, Anaish. In retaliation, the Hualapais cut Prescott's communication lines and killed several squatters. In March 1866 two white men murdered a prospector at the Willows east of Beale's Springs but cast blame on the Hualapais. Anglos retaliated by killing the respected Chief Wauba Yuma of the Yavapai Fighters and his two sons specifically to provoke hostilities and thereby increase their profits by keeping the army in the area (Dobyns and Euler 1976: 36). As Hualapai elder Kate Crozier remembered of this time, "The white man came from the east. They stole horses. . . . The white man just killed off the Indians living here and their relatives too. They went on killing, this I saw" (Hinton and Watahomigie 1984:93). John Williams, a Yavapai elder, also spoke of the horrors encountered by his people:

When the white people come, lots of Yavapai get killed. This here is our home. But when the white people come, they take it away from us. They start mining out of these big rocks. We know nothing about these things. White people come here and they get gold, they get silver, they get copper, they get iron, they get lead out of these rocks. If the white people would not kill the Indians, it would be all right. But they kill them, and I don't know why they do that. Before the white people live here, we have no trouble at all. At that time we all stay together, not like now, all separate. Now there are few of us left and that is why we are scattered out. (Hinton and Watahomigie 1984:187)

But the Upland Yumans did not give up their land without a fight. Bands retaliated and the "Hualapai Wars" began (Dobyns and Euler 1971). As John C. Dunn, the Indian agent in Prescott, reported to his superior on May 23, 1865:

We have a war waged upon us by the Yavapais, Hualapais, and Apache Mohaves, which has been brought on by the wanton and cruel aggressions of not only the settlers, but by the troops placed here for protection and peace. . . . Some time last January Captain Thompson, now in command at Fort Whipple, went out on a scout into the Hualapais country, with one George Cooler as guide, and surprised a ranchería of Yavapais or Apache Mohaves, killing 28 men, women, and children, among the number Hoseckrus, a Yavapai chief, who has ever shown himself a true friend of the whites. Not more than 10 days previous to this attack, some of these same Indians were at work for me on the toll road at very small wages, and were friendly and trusting. They herded our stock, and when it strayed they searched and brought it back. The attack upon them by the troops was an unprovoked outrage and murder. (U.S. Senate 1936:42)

The Pais were tenacious fighters, and Colonel William Price noted that "officers from Prescott say they would prefer fighting five Apaches to one Hualapai" (Dobyns and Euler 1971:46). After a surprise attack on Cherum's camp, Captain Young confessed, "We had been fighting our best for one hour and twenty-five minutes when Indians made their appearance on both flanks and I withdrew slowly from that vestibule of death" (U.S. Senate 1936:49). Hualapai leaders gained great renown for their brilliant tactics. But while they won pitched battles (Dobyns and Euler 1960), the army's scorched-earth tactics eventually prevailed, and bands retreated to the mountains and canyons. There the army slowly destroyed one group after another. U.S. cavalry detachments burned at least sixty-eight rancherías and innumerable fields, destroyed tools and equipment, and killed nearly a fourth of all the Hualapais.

After the last Hualapais surrendered in 1869 following a whooping-cough epidemic, a one-square-mile reservation was established for them at Camp Beale Spring, ostensibly to protect them from miners. As the area was too small to support more than one or two families, bands left and tried to avoid further contact with Anglos, but this was nearly impossible. The Bureau of Indian Affairs (BIA), following their policy of removing conquered peoples from ancestral lands, instructed the army to move the Hualapais to the Colorado River Reservation in 1874. The Havasupai and Peach Springs Canyon bands, who had been less directly involved in the war, hid in the Grand Canyon and its side canyons.

The Hualapais suffered greatly on their Long Walk to La Paz. There they were treated as prisoners of war. Government rations were in short supply, and the promised cattle never arrived. The Hualapais did not know how to farm on a floodplain, and the BIA-designed irrigation canals did not work. Devastated by disease and hunger, nearly half died. Brokenhearted, the Hualapais fled in April 1875, only to discover that Anglo American settlers had appropriated their springs and that cattle were grazing on their hunting and gathering lands. Some tried to resume their old economic patterns, but settlers demanded their removal. Others became miners, and because they were such good workers, mine operators fought for a permanent Hualapai reservation near the mines. An executive order established a Hualapai reservation of 997,045 acres along the south rim of the Grand Canyon in 1883. The new reservation represented about one-tenth of their original territory.

The Yavapais met the same fate as the Hualapais. In 1871 approximately 750 Yavapais and Apaches were impounded by executive order on a large reservation covering most of the Verde Valley. But in 1872 another executive order canceled the reservation, and the Yavapais were ordered to move. When they resisted and resumed raiding in order to survive, the army, under General George Crook, began a scorched-earth campaign. The ensuing army raids were notorious, even among army personnel, and they are still recounted as dark days in Yavapai histories. One of the worst was the 1872 Skeleton Cave Massacre, in which the military brutally murdered an entire band—seventy-six Yavapai men, women, and children.

Finally overwhelmed by Crook's soldiers, most of the Yavapais surrendered in April 1873. More were captured the next year and were incarcerated on the Verde River Reservation. The government promised them a ten-mile-long canal to irrigate 2,000 acres. Despite a year of squabbling between military and civilian officials, the Yavapais managed to excavate an irrigation ditch with wooden sticks and produce successful harvests. Unfortunately, a group of Tucson contractors who supplied Indian reservations were alarmed by their growing self-sufficiency and pressed the government to relocate them. In 1875, as prisoners of war, 1,500 individuals were marched 180 miles to the San Carlos Apache Reservation near Globe, Arizona (Corbusier 1969). On the midwinter March of Tears, while soldiers rode ponies, more than a hundred Yavapais died on the eight-day trek. "With no time for burial, they were left lying along the way" (Barnett 1968:3).

Unlike other Upland Yuman groups, the army left the Havasupais alone; they were not forced to relocate. Interaction with the few Anglo ranchers and miners who ventured into the area was rarely violent. As a result, the Havasupais were not defeated, but they did suffer losses of their land to Mormons and Navajos. In the late 1870s, government officials, fearing that their central springs would be encroached upon, decided to establish a reservation. Unfortunately, officials had little understanding of how the Havasupais actually earned a living; on June 8, 1880, the "Suppai" were forced off the plateau and restricted to a 518-acre parcel of bottomland in Cataract Canyon, a process that separated them from all their hunting, gathering, and winter garden plots.

The Reservation Period

Reservation lands remained in flux for many years due to demands from settlers, ranchers, miners, and the railroad. Even the conservation movement imposed new hardships. In 1908 the Grand Canyon was set aside as a national monument and game preserve without considering Havasupai rights and needs. This conservation of flora, fauna, and landscape undermined the Havasupais' economy and cost them 90 percent of their land. They no longer had skins to trade, and they lost woodcutting areas and control of the watershed. Confined to their summer farming lands and cut off from winter hunting grounds, the Havasupais had no way to replace their drastically curtailed food supply even though they intensified their agriculture. This situation created extreme poverty. The Havasupais led a much more precarious life than they had in the early nineteenth century. Nor did the government make up the shortages, for entry into the canyon was difficult, so government services were limited. The population dropped between 1898 and 1919 due to a series of epidemics and disastrous floods. In 1920 there were only fifty households in the canyon, totaling 175 individuals. The situation began to improve only when the Havasupais gained enough concessions to start their own cattle herd and obtained permits to use land in the Coconino National Forest.

The Hualapais faced a similar situation. Their reservation consisted of grassland and forest, little of which was tillable, so they could raise few crops. Most Pais settled at Peach Springs, the main source of permanent water on their remaining lands. Some Hualapais turned to stock raising, but Anglo ranchers, whose private inholdings dotted the reservation, continued to preempt key springs. As one elder reported to the Indian Claims Commission in 1950:

> When these cattlemen came in and put their horses and cattle on this water, on this good country range, they told us that it was theirs, that is their water, that is their land. . . . These cowmen just had their guns ready, you know, and we were scared of them. And that is why we don't live close to these waters anymore. (U.S. Indian Claims Commission 1950:75)

Under these conditions, the reservation could support only about a fifth of the tribe. Most worked in the mines or for Anglo ranchers, joined the army, or moved to railroad towns and worked as low-paid unskilled laborers—cutting hay, hauling timber, working as domestic servants—occupations at the bottom of the Anglo socioeconomic hierarchy. By 1887, one in four of the Pais had jobs. "I worked for a mine, pushing a wheelbarrow or cutting wood for twenty-five or fifty cents a day," a Pai named Kuni said. "When I had fifteen to twenty dollars I would bring flour and sugar back to my father" (Kroeber 1935:207). Many Hualapais were reduced to begging and prostitution, but local whites were not impressed by their plight. A common attitude of settlers is reflected in this comment from the *Mohave County Miner* of October 8, 1887: "The Wallapai Indians complain of the quality of the flour served out to them by the government, and say it is full of weevils and has an intensely bitter taste. A plentiful supply of arsenic mixed with it would disguise the bitter taste."

All but powerless in the face of government restrictions and facing open hostility from settlers, Hualapais looked for ways to reclaim their culture. They turned to a revitalization and a millenarian movement, the Ghost Dance, as a way of returning to aboriginal conditions. The Ghost Dance was founded by Wovoka, a Northern Paiute mystic who prophesied that certain rituals, correctly performed, would restore the old Indian ways and resurrect the dead. The Ghost Dance was a message of Indian hope that stressed peace and emphasized the need to learn new skills and work diligently in order to prosper in this life. In 1889 Panamita, a Paiute missionary, and Doinhu'uka (Jeff), a respected Hualapai shaman who received his calling in a dream while visiting the Paiute, introduced the movement to the Pais. Five hundred participants moved in a traditional round dance around a tall pole at Grass Springs. Several fell into trances and had visions of a world without white oppressors. The ritual lasted several days, until food stores were exhausted. Over the course of subsequent dances, the Hualapais modified Wovoka's message: the recovery of land and the return of game and health, rather than a return to the old ways, came to be at the heart of the movement (Dobyns and Euler 1976).

When white settlers heard of the activities, they concluded that the Ghost Dance's appearance presaged new Indian uprisings. As bands disappeared, settlers in Wickenburg thought that they were preparing for war and sent J. H. Johnson and John Kolar to watch the ceremonies. In the *Mohave County Miner* of September 28, 1889, they reported that the five hundred Hualapais attending the first dance were peaceful, although their rhetoric was not. They realized that the adherents would resort only to magical means to attain their ends. When the promised new order did not materialize, the Hualapais abandoned the Ghost Dance in 1895. The failure of the Ghost Dance to remove Anglos may have dealt a blow to traditional beliefs for some individuals.

Reservation life drastically changed the life of all Upland Yumans, but the most affected were the Yavapais, who were mistakenly and unfortunately treated as Western Apaches. Since coming to the Southwest, Anglo American explorers and army officials, like the Spanish and Mexicans who had preceded them, had confused the southeastern Yavapais and the Tonto Apaches, referring to both groups as Mohave-Apaches or Yuma-Apaches. While incarcerated at San Carlos, in the homeland of the San Carlos Apaches, whom the Southeastern Yavapais considered to be their enemies (Gifford 1932:180), the Yavapai population quickly declined from 1,500 in 1875 to only 500 in 1905. Nevertheless, Yavapais resiliently tried new ways to survive. They enlisted as scouts in the U.S. Army, became miners, cowboys, or policemen, or worked as skilled or unskilled laborers. But without their land, their tribal identity was threatened. Yavapai leaders began to work toward the establishment of separate reservations on their ancestral lands.

In 1898 the Yavapais applied to the federal government for permission to return home. The permission was granted, but no provision for land was made, so in the early 1900s families quietly began to drift away from the San Carlos reservation. Families and bands, along with some Tonto Apache relatives, slowly dispersed throughout much of the old lands. One group moved to a site near Phoenix, where the Fort McDowell Reservation was established in 1903. Others went back to Camp Verde to farm on the forty-acre abandoned military reservation. By 1912 several men were working in the copper mines and smelter at Clarkdale, while others were working on road and construction projects such as the hydroelectric plant at Fossil Creek. Don Mitchell, a Yavapai elder, noted that his family earned money by cutting wood, doing laundry, and selling baskets. Others subsisted mostly by hunting game, collecting wild plants, and trading game for corn (Keller and Stein 1985:10).

When Anglo-American settlers in the Verde Valley began to complain that the Yavapais were reestablishing their traditional lifestyle, small parcels of noncontiguous land were "set aside" as the Camp Verde Reservation: Camp Verde, Middle Verde, and Rimrock. Other small reservations were established at abandoned forts—minuscule sections of the Yavapais' original territory, each of which could only support small numbers of people. Four hundred individuals settled at Fort McDowell, and another 500 in the Camp Verde area. While the population slowly increased, devastating setbacks still occurred. More than half of the people at Fort McDowell died in the influenza epidemic of 1918. Another reservation was established in Prescott in June 1935, when seventy-five acres were transferred from the Veterans Administration (at old Fort Whipple) to the Department of the Interior in trust for the Yavapais. But many Yavapais had to live off-reservation because few families could earn their living from farming on any of the reservations' lands. The slowdown and finally the closing of the mines in central Arizona during the 1930s and 1940s greatly affected Yavapai and Hualapai workers. As a consequence, more people returned to the reservations. In 1969, sixty acres near Clarkdale were set aside as the newest Yavapai reservation.

Education and Bilingualism as the Road to Self-Determination

Treaties commonly stipulated that the federal government would educate American Indian children. A teacher and farmer came to the Havasupai reservation in the 1890s and brought new tools, new ideas, and the English language. Many Pais decided that education was necessary to reestablish their way of life, so they began to send their children to BIA schools. Kuni remembered Walapai Charlie saying, "When you send your

Qua-tha, a Yavapai scout, in about 1882. (Arizona Historical Society)

children to that school, they learn something to help you; they
will learn English; they will learn to read and write and to interpret" (Kroeber 1935:107). Due to the small, dispersed populations, however, many children had to attend boarding schools,
generally against the wishes of parents. Children were given
Western-style clothing and were taught to be farmers and wage
workers who would no longer be a "burden" to the government.
But boarding schools created hardships at home because children could not learn about their customs and lands. Hualapais
and Havasupais wanted schools on their own lands. Today an
elementary school can be found in Havasu Canyon and middle
and high schools in Williams, Kingman, Truxon Canyon, Selig-
man, and Phoenix.

Education had unexpected results; it became a weapon in the
fight for self-determination. This can be seen in the successes of
Dr. Carlos Montezuma, a Yavapai physician, who raised a voice

for Indian land and water rights that was heard across the country. At a time when American Indians had virtually no political rights, Montezuma fought for equality for all American Indians, with immediate results for the Yavapais. In 1906 the BIA recommended that no funds be spent on the irrigation system at Fort McDowell. Yavapai farmers had labored for years to develop and maintain this system, which was periodically washed out by the flooding of the Verde River. The BIA proposed relocating Fort McDowell residents to the Salt River Pima-Maricopa Reservation. Of course, this meant that the Yavapais would live without water rights on the lands of their traditional enemies, the Pimas (Akimel O'odham). Montezuma learned of the plan and was able to mobilize national leaders. The Yavapais remained at Fort McDowell. Over the next thirty years, the Yavapais were able to defeat various other relocation attempts through similar political activism.

The Yavapai effort to fight discrimination through the courts was another offshoot of this educational effort. While the federal government had passed a law in 1924 that finally recognized American Indians as citizens who had the right to vote in federal and state elections, Arizona instituted laws that denied them a voice in the plebiscite. This injustice came to an end in 1948, when two Fort McDowell Yavapais, Frank Harrison and Harry Austin, filed suit and won (*Harrison v. Laveen*, 67 ARIZ. 337, 196 P2d 456). All Arizona Indians benefited from their initiative.

In the late 1970s, Upland Yumans used education to tackle another problem: language retention. The Hualapais developed a Pai literacy program, the Hualapai Bilingual/Bicultural Education Program, based on their belief that Indian cultures will only flourish if their languages are actively used. Since the founding of the program in 1976, program personnel have published bilingual educational materials (oral history texts, newsletters, and pamphlets) and have distributed them to the community. This award-winning program has served as a model for native peoples throughout the United States.

The Hualapais have also begun to utilize the techniques of formal linguistics. Led by Lucille Watahomigie, educators attend linguistic conferences and engage in new collaborations with Anglo scholars. Pais have also become concerned with what is

LUCILLE J. WATAHOMIGIE
SCHOLAR AND EDUCATOR

Nancy J. Parezo

Lucille Watahomigie was raised at Milkweed, Arizona, on the Hualapai reservation in a family that included her grandparents and other relatives. Her childhood was traditional since her family based their living on agriculture, hunting, and gathering wild foods. Her life combines this traditional world with that of Anglo-style scholarship. After earning a master's degree in education from the University of Arizona, she became the associate director of the university's Teacher Education Program for Indian Students.

In 1976 Watahomigie returned to the Hualapai reservation to serve as the director of the newly established Hualapai Bilingual/Bicultural Education Program. Under her direction, this program became one of the most successful in the country, receiving numerous grants and awards. Watahomigie has shared her experience with many native peoples in the United States and Canada, who have come to see how the program works.

Watahomigie has been responsible for the collection, transcription, and translation of numerous stories and texts. She feels this is important information for her people:

> The oral tradition of the Hualapai transmits the language, history, values, and mores of the culture. Oral tradition teaches the way things are to be: sharing, helping one another, living in harmony with nature, and the basic philosophy of the culture. Oral tradition is the conveyor of memories, passed on from generation to generation, memories of the land once occupied by the Hualapai, contacts with other tribes, their history; it tells the people who they are today on the basis of who they were yesterday. (Watahomigie 1984: 11)

As a native speaker and a trained linguist, Watahomigie has been an important force in Yuman linguistics. She has initiated and run a series of summer workshops on Yuman languages. Funded under Title VII and held at Arizona State University, the workshops have recently expanded in scope to include the training of other native Southwestern speakers in linguistic methodology. She has also had a fruitful collaboration with other linguists. One of the more important of these endeavors has resulted in the *Hualapai Reference Grammar* (Watahomigie, Bender, and Yamamoto 1982) and *Spirit Mountain* (Hinton and Watahomigie 1984), an anthology of Yuman stories and songs.

being published about them and how this information is presented. The Hualapais are increasingly collecting their own data
and turning to linguists to help train them to undertake their
own studies. The collective authorship that has resulted has
greatly improved southwestern linguistics.

Moving Forward

The Upland Yumans, who are among the smallest tribes in Arizona, are still waging a fierce battle to maintain their rights to
traditional lands. Havasupais, Hualapais, and Yavapais have to
be continually vigilant to retain even the remaining fraction of
their original territories because attacks on their lands from
ranchers, railroad companies, the government, and environmentalists have continued into the 1980s. Now the battle is waged in
the courtroom and the halls of Congress, and there are indications of success.

In 1947 the Hualapai reservation was consolidated when
500,000 acres of odd-numbered sections previously held by the
Santa Fe Railroad were placed in trust for them. In a 1960s land-
claims case, the Hualapais received $2.9 million in compensation
for lands taken from them in 1883. The Havasupais likewise
have worked vigorously to regain their plateau lands. In 1972
the Havasupai Tribal Council declared that "this land is part of
us," and in 1973 it told the world that "four hundred of us are
taking on the machinery of the United States government, and
we intend to win." With the support of Representative Morris
Udall and Senator Barry Goldwater, the Havasupais survived
bitter attacks from the National Park Service and the Sierra
Club, who wanted the land declared a wildlife area. On January
3, 1975, President Gerald Ford signed a bill that restored 160,000
acres of land to the Havasupais. An additional 95,000 acres of
Grand Canyon National Park was allocated for their traditional
use. As a compromise, the Havasupais agreed to admit a limited
number of tourists—backpackers and those on horses. They still
refuse to build jeep trails or tramways into the canyon despite
many proposals from outsiders. Tourism accounts for roughly
half of the Havasupai Tribe's income, but it is tourism on the
Havasupais' terms.

The Yavapais have faced similar attacks on their reservation lands. Phoenix's ever-growing need for water has continually placed Yavapai lands in jeopardy. In 1968, after decades of court battles and legislative struggles, Congress approved the Central Arizona Project. Designed to bring Colorado River water to central and southern Arizona, the project included construction of Orme Dam, which would have flooded nearly two-thirds of the Fort McDowell Reservation and required relocation. The affected acres consisted of all the fertile bottomlands used for farming, cattle grazing, woodcutting, and housing. The Yavapais were not consulted about this project; they were only told they would receive relocation funds and compensation for their lands under the principle of eminent domain. The threat of the dam halted economic development and federal aid for housing, health, and agricultural improvements.

In an informal referendum at Fort McDowell in 1966, the Yavapais voted against the dam, but no one listened. In 1968, Congress passed the Colorado River Basin Project Construction Act (Public Law 90-537), which included the Orme Dam. Yavapais continued to protest, but it was more than four years before Bureau of Reclamation personnel met with the community. As John Williams, a Yavapai elder, repeatedly told government officials, "I can't trade my land with money. I don't care how much money you give me. I won't take it. That's the only good place, right where I am." For several more years the government ignored Yavapai protests. In 1981 Yavapai elders and environmentalists, who opposed the dam because it would have flooded the nesting sites of endangered bald eagles, walked the thirty-mile Trail of Tears. Finally, after cost estimates skyrocketed, the government canceled plans for the dam.

The Havasupais, Hualapais, and Yavapais have retained remarkably consistent concepts of themselves as peoples with rights to their lands despite years of rapid change under U.S. rule. They continue to fight for their lands as a visible symbol of their right to self-determination because they believe it is the land which makes them who and what they are. For small populations, sometimes living on the outskirts of expanding cities or on lands that environmentalists want protected, these fights will be difficult, for the United States is a country where political influence often correlates with the size of the group and the amount of money that can be given to political campaigns. But as the Hualapais have recently shown in their stand against the federal government's confiscation of gaming equipment, the Pais will continue to stand proudly and succeed even against great odds. In the words of Wilfred R. Watoname, Sr.,

> Our Hualapai Tribe had great leaders who laid a foundation upon which we have been able to build and to maintain through the years our pride, dignity, culture, and honesty, which is a measure of our past leaders. We have not been stripped of these qualities and will always have them for we are the First Americans. (Dobyns and Euler 1976:vi)

THE O'ODHAM

(PIMAS AND PAPAGOS)

THE WORLD WOULD

BURN WITHOUT RAIN

Thomas E. Sheridan

When Jesuit missionary Eusebio Francisco Kino embarked upon the missionization of the Pimería Alta (Upper Pima country) in 1687, he set off across a vast desert dominated by people who called themselves the O'odham. Kino and his companions called the O'odham the Pima Altas, or Upper Pimas, to distinguish them from the Pima Bajos, or Lower Pimas, living in southern Sonora. But they referred to them by a number of regional names as well: Hímeri for the groups living along the Dolores and Magdalena Rivers in northern Sonora, Soba for the people along the Río Concepción, Sobaipuri for the O'odham of the Santa Cruz and San Pedro Valleys, Gileño for the river dwellers in central Arizona, and Papago for the bands who inhabited the riverless basin-and-range country between the Santa Cruz and Colorado Rivers. Those names did not capture the ways in which the O'odham defined themselves, but they did reflect the Europeans' awareness of the immensity of the Pimería Alta and the diversity of the people who lived there.

During the next three hundred years, many of those groups disappeared, victims of disease, displacement, and miscegenation. The Apaches drove the Sobaipuris out of the San Pedro Valley, and Spaniards and Mexicans forced the Sobas and Hímeris away from the rivers of northern Sonora. In Arizona, however, thousands of O'odham survived into the twentieth century. By then they were drawing one major distinction among themselves. Those living along the Salt and Gila Rivers were the Akimel O'odham, or River People, and those in the western

deserts were the Tohono O'odham, or Desert People. Being O'odham meant many different things—the language people spoke,
the way they treated their neighbors, the conceptions of the
universe they held. But another distinguishing characteristic was
water—whether it flowed year-round past a village or whether
the people had to move from mountain spring to valley arroyo in
search of it during the desert year. Water made the crops grow
and the cactus fruits ripen. Water shaped the size of community
and the activities of the seasons. The O'odham therefore sang
about water, dreamed about water, and learned all the ways to
utilize whatever water they found in a dry land. According to one
song, I'itoi, the elusive Elder Brother of the Tohono O'odham,
told the people that the world would burn without rain. In the
Sonoran Desert the truth of that statement was more than meta-
phorical.

O'odham and O'dami

Some archaeologists believe that the O'odham are among the
most ancient of southwestern peoples, and have occupied the
Sonoran Desert long before the rise of the Hohokam about A.D.
200. Others, like Emil Haury, contend that they are descendants
of the Hohokam, that a Hohokam-Pima continuum extends
from prehistoric times to the present. A handful of scholars like
anthropologist Bernard Fontana and ethnobiologist Amadeo
Rea, on the other hand, argue that the O'odham are relative
newcomers who moved into the region as Hohokam civilization
was collapsing in the 1400s. According to Rea's scenario, the
O'odham migrated to southern Arizona from the desert lowlands of southern Sonora sometime between 1400 and the arrival of Kino in 1694. One line of evidence Rea cites is eth-
nobotanical: the ways in which the Pima Bajos classify plants
are, he says, "amazingly parallel" to the plant ethnotaxa (classi-
fications) of the Gila Pimas in central Arizona. But Rea analyzes
different versions of the Gila Pima emergence myth as well,
including the one recorded by anthropologist Frank Russell in
the early 1900s. In Rea's words, "While told in epic style, it
would be hard not to read these as historic events, supernatu-

rally embellished in much the same manner as the Hebrew Exodus" (Rea ms.).

Frank Russell's version goes as follows. After being killed by the people he helped create, Elder Brother revived himself and sank to the other side of the earth to find the people who had fled there to escape a great flood. Elder Brother told them of his mistreatment and asked them to help him reclaim the land from those who had murdered him. The ancestors of the Pimas agreed, crawling through a hole in the earth bored by Gopher. They emerged somewhere to the east of Casa Grande and made war on the people living in the great pueblos along the Gila and Salt Rivers. As they approached Casa Grande, they sang:

Yonder stands the doomed habitation.
  About the pueblo runs its frightened chieftain
In yellow garment with hand-print decoration.
                                        RUSSELL 1975:281

After overrunning Casa Grande, the Pimas accompanying Elder Brother conquered seven more pueblos, defeating the *siivañi* (chieftains with magical powers) who ruled them. Many of the inhabitants were killed; others ran down the Gila River to the Colorado, eventually turning northeast to settle in the pueblos of the Rio Grande. According to Joe Giff, an elderly Pima steeped in traditional lore, the ones who fled were "the *vipishad*, the small people, builders of the ruins," who "inhabited the earth before the Pima." The Pimas, in contrast, were the *vupushkam*, "the ones who came up." Giff believed that the vipishad were the Hopis (Rea ms.). Interestingly enough, the Hopis themselves contend that some of their clans originated among the Hohokam.

We may never know which of these hypotheses is correct— whether the O'odham lived in the Sonoran Desert before the Hohokam, whether they descended from the Hohokam, or whether they drove the Hohokam from the region. Regardless of when the Pimas settled in Arizona, however, we do know this: The O'odham belong to a long, broken arc of Piman speakers stretching from the Salt River Valley near Phoenix to the mountains of northern Jalisco. Three centuries ago, when Spaniards began moving into southern Arizona, an O'odham along the

Gila River could have walked south for more than a thousand miles and communicated in his or her language with the Lower Pimas of southern Sonora, the Tepehuanes of Chihuahua and Durango, and the Tepecanos of Jalisco. Even the names these groups call themselves are nearly identical. Lower Pimas are O'odham, and the Tepehuanes are O'dami. Some archaeologists who believe that the Hohokam were Piman speakers argue that Mesoamerican influences in the Southwest diffused northward along this "Tepiman" corridor. Whether or not that occurred, the similarity of the names reveals the cultural as well as linguistic unity of the so-called Tepiman connection from the southern Sierra Madre to the Sonoran Desert.

Rain and Saguaro Wine

The Mountain Pimas and the Tepehuanes occupy high, pine-fringed valleys in the Sierra Madre of northern Mexico. They plant corn, beans, squash, and oats at elevations of 5,000 to 8,000 feet, and their lives are shaped by the same contrast between mountain and canyon as are their neighbors, the Tarahumaras, who border them to the east and north. The Tohono and Akimel O'odham, on the other hand, are desert people. Even those who used to live along rivers like the Gila, the San Pedro, and the Santa Cruz relied, at least in part, on desert plants and animals. The O'odham New Year, in fact, begins not in January but in late June or early July, when the fruits of the giant saguaro ripen. The O'odham gather those fruits with long poles made of saguaro ribs (see pls. 22 and 23). Then they boil down the juicy red pulp into syrup (see pl. 24). The saguaro harvest marks the onset of the most abundant time of year—a time when cactus fruits and mesquite pods are ready to be eaten, when wild greens push through the desert soil, when fields can be planted. It is the season of the summer rains.

A legend of the Tohono O'odham reveals how deeply the relation between the saguaro harvest and the coming of the rains has penetrated O'odham culture. It says that, long ago, a dust devil attacked the daughter of a powerful man. Enraged, her father persuaded his neighbors to drive the wind away. But when Wind departed, he took along his blind friend, Rain. For four years, no rain fell. Everything withered. All living things suffered.

When they could stand the heat and thirst no longer, the O'odham asked Coyote to find Wind and Rain and beg them to return. Coyote searched for four days without success. The O'odham then sent Bear and Buzzard, who also failed. Finally, in desperation the people turned to Hummingbird. Tying some of his down on a stick, Hummingbird flew across the earth until he saw the down stir.

Hummingbird then followed the breeze to a cave where Wind and Rain were sleeping. To wake them, he put a hot coal from their fire against their backs. Wind and Rain remembered their rejection by the O'odham, however, and at first they were unwilling to return. Wind said, "Tell our relatives that if they want us, they must sing for us for four nights. We'll return when they finish the ceremony."

Hummingbird went back to the O'odham with Wind's demand. The old men consulted with one another and decided that they needed to make Wind forget their harsh words. To do so, they made *nawait*—saguaro wine—and performed the saguaro wine feast (Saxton and Saxton 1973:317–40).

Ceremony and Survival in the Desert

The saguaro wine feast is the first public ritual of the Tohono O'odham year. Individual families may gather saguaro fruits and turn them into syrup, but the syrup is fermented in the communal roundhouse (*o:las ki:*) of each village. Two nights before the fermentation is complete, the people of the village gather to sing and dance around a fire in the middle of the ceremonial grounds. Those grounds—bare and hard packed—are always located east of the roundhouse and the ramada (*watto*) in front of it (Underhill et al. 1979:19–35).

After the dawn of the second night, when the wine is tasted, several men run to three neighboring villages to invite them to the wine feast. There they deliver a ritual oration, which is known as the running speech. Men sitting in front of the roundhouses in the invited villages respond with another oration called the return running. Significantly, both speeches begin by describing the desolation of the land without rain.

Later that morning, the invited villagers arrive at the host village. There they are greeted with a seating speech (*dadṣpa ñiok*), and then the sit-and-drink ceremony (*dahiwak i:*) begins. The ceremony is complex and includes both sermons and ritual orations, as well as the consumption of the saguaro wine. It ends with one of the most beautiful of all O'odham orations, the mockingbird speech (*ṣu:gaj ñiok*).

The mockingbird speech vividly reveals the connection between drinking the wine and releasing the rain. Mimicking the O'odham just as he mimics the songs of other birds, Mockingbird swallows "medicine man's liquid, rainmaker's liquid," and becomes joyously drunk. His celebration is heard and understood by Wind, who brings dust storms, lightning, thunder, and finally an enormous rainstorm. When the storm is over:

> The lying earth was beautifully wet and
>   finished
> On top of that came out various seeded
>   things, *opon*, su'uwat, da:pk [all
>   wild greens or grasses], and every
>   kind of cactus, well did they ripen.
> It was this that was our last wish and it
>   happened
> As you will see.
>
> UNDERHILL et al. 1979:35

Tohono O'odham Agriculture: The Importance of the Summer Rains

In cadence and poetic phrase, the mockingbird speech conjures up that most important of environmental events: the onset of the summer rainy season, when Wind and Rain return. The Tohono O'odham occupy one of the driest stretches of the Sonoran Desert. Much of the region averages less than five inches of rain a year, and more than half that moisture falls in violent, highly localized storms during July, August, and September.

A Tohono O'odham calendar stick from Sil Nakya, collected in 1939 from keeper José Mara. The sticks were personal mnemonic devices, with years divided by lines. This stick begins in 1841 and ends in 1939. The marks denote important events ranging from *wi:gida* ceremonies to deaths from automobile accidents. (Photograph by Helga Teiwes, Arizona State Museum)

These storms lash the desert with blinding sheets of rain. The rain falls too hard and too fast for the desert soil to absorb much of it, so a thousand rills pour into normally dry arroyos that fill with terrifying suddenness. Surging with brown currents of runoff and the nitrogen-rich detritus of the desert floor, the arroyos often overflow, soaking the alluvial soil along their banks. Before pump-powered wells were drilled in the early twentieth century, arroyo floodwaters provided the Tohono O'odham with the only source of moisture for their fields.

Gathering saguaro fruits and making saguaro wine therefore are as much a part of the agricultural cycle as planting seeds in the wet earth. By drinking saguaro wine, the Tohono O'odham call Wind, who brings Rain. Coded in ritual speech and song, a web of associations links a whole series of activities that may appear unrelated to outsiders. But the relations are there, and they all rest on the need for summer rain.

Most early Anglo American visitors were unaware of that web and painted a bleak picture of the Papaguería. According to D. D. Gaillard, a soldier who helped survey the international boundary in the early 1890s, "The entire region is a hopeless desert, and few if any Americans reside in it. It is unadapted for agriculture, yet when the July rains commence, the Indians forsake their rancherías and hasten to their *temporales* [floodwater fields] where they plant crops of corn, pumpkins, melons, squashes, etc." (Gaillard 1894:293).

But even if they sneered at the landscape, a few of those Anglos expressed a grudging respect for the ability of the O'odham to survive there. As Indian agent E. A. Howard sardonically commented in 1887, "Place the same number of whites on a barren, sandy desert such as they live on, and tell them to subsist there; the probability is that in two years they would become extinct" (Howard 1887:47).

To the Tohono O'odham, however, their desert was anything but barren or hopeless. On the contrary, it was a place of beauty and life, graced by the poetry of sunlight, storm clouds, and magical creatures like Turtle, Rattlesnake, and "my shining helper," the trickster Coyote. But people had to know where and when to look for that life, so the O'odham developed a society that was fluid and flexible enough to adapt to the desert rather than trying to make the desert adapt to them. Following Ruth Underhill, the pioneer ethnographer of the Tohono O'odham, anthropologist Bernard Fontana called historic Papagos the Two Villagers (Fontana et al. 1983a). During the winter months, most O'odham lived near permanent springs in the foothills of the mountain ranges that rise out of the desert. During the summer, they moved down into the alluvial valleys, where they planted small plots of corn, beans, squash, and devil's claw along desert washes. Gaillard captured the frenetic vitality of those temporary summer communities:

> At many places on the desert where fertile land can be found near water holes, or convenient to their artificially constructed dams, they establish "Temporales" and fence in with mesquite brush small fields, to which they promptly repair when the first summer rain falls. Where silence reigned before, all is now full of activity. Houses and fences are repaired; irrigation ditches are put in order; new dams built or old ones repaired; and often within 24 hours after the first drop of rain falls the entire crop of melons, pumpkins, squashes, beans, and Indian corn are planted. (Gaillard 1896:601)

Also manifesting O'odham versatility were the crops themselves. Selected over many generations by observant O'odham farmers, the cultigens were extraordinarily well adapted to heat and drought. Under arid conditions, for example, the small tepary bean (*Phaseolus acutifolius*) consistently outyields the larger, more common pinto bean. Ethnobotanist Gary Nabhan and his colleagues also discovered that teparies grown in Tohono O'od-

A Tohono O'odham woman's carrying basket (*giho*) made from long, dried saguaro ribs, twisted agave fiber netting, and a plaited sotol back pad and tump strap. Such baskets were used for carrying firewood, water ollas, and other burdens on the back. This basket was made sometime between 1865 and 1890. (Photograph by Helga Teiwes, Arizona State Museum)

ham temporales have a higher protein content than those raised in irrigated fields. They hypothesize that when tepary plants are "starved" for water, they devote more energy to producing seeds than foliage. Teparies were important enough to the Tohono O'odham to give them their nickname—Bawi Kuadam, or Tepary Eaters. Some linguists even believe that the term Papago derives from the nickname itself.

At this point, it is important to distinguish between the runoff (*temporal*) agriculture of the Tohono O'odham and the irrigation agriculture of their Pima neighbors. The Tohono O'odham constructed waterworks along the arroyos of the Papaguería, which included crude tanks (*charcos*) to impound water in the arroyos and brush dikes and shallow ditches to channel runoff

THE WI:GIDA CEREMONY

Thomas E. Sheridan

On the evening of November 26, 1945, archaeologist Julian Hayden left Tucson and drove west across the Tohono O'odham Reservation. He was accompanied by Juan Xavier, a Tohono O'odham, and Xavier's wife, Gwyneth Harrington, an anthropologist. They reached the village of Kakaij Mek (Santa Rosa, south of the wash) around midnight and followed wagon ruts to the community of Ge Aji. There, Hayden and his friends "saw the semi-circle of campfires, the covered wagons and buckboards drawn up and between them, the dance enclosure with its glowing column of smoke and sparks eddying upward" (Hayden 1987:296). Although they did not know it then, the three visitors were witnessing the last winter *wi:gida* performed by the O'odham.

The wi:gida is one of the Southwest's most intriguing mysteries—a ritual that offers tantalizing glimpses into the pre-Columbian past not just of the O'odham but of other southwestern peoples as well. The term derives from the O'odham word *wi:gĭ* (bird down or eagle down). Considered by most scholars to be a harvest feast, the winter wi:gida was supposed to be performed every four years in Ge Aji. A summer wi:gida continues to be held each year in the village of Quitovac just south of the international border.

What makes the wi:gida so seductive, however, is its echoes of past relationships among peoples now widely separated by time and space. The most elaborate of all O'odham ceremonies, the wi:gida centers on masked dancers with headdresses of turkey tail feathers—characteristics that appear to be of Puebloan origin. The anthropologist Ruth Underhill, in fact, argued that the ritual was not a harvest festival at all. In her words, the purpose of the wi:gida was to "'keep the world in order' and prevent a flood. Keeping the world in order was a function of the Pueblo solstice ceremonies, and, indeed, the respective dates of the prayerstick festival comes as near to summer and winter solstices as they do to harvest" (Underhill 1946: 154).

How could a ceremony of Puebloan origin have taken root in one of the driest parts of the Sonoran Desert? The Tohono O'odham believe that the wi:gida was given to them by a man called Naviju, which means friend (*navoj*) in O'odham (see pl. 21). Naviju passed through the Papaguería and wandered into Mexico but returned to the village of Agam north of Ge Aji. There he taught the O'odham how to design the costumes and perform the dances of the wi:gida, which observes the founding of the Tohono O'odham's original five villages—Ge Aji, Kakaij Mek, Ge Oidag, Ak-cin, and Añgam. The Nanviju dancers represented Naviju, and many of them dreamed new songs each time the winter ceremony was presented. According to Hayden, "The songs, which came in dreams, appear to be concerned quite naturally with the harvest and with natural phenomena, as well as with the landmarks of Papaguería which have legendary and supernatural associations" (Hayden 1987:295). "Keeping the world in order" meant good harvests and rain, just as it did among the Hopis and other Western Pueblo peoples.

In Hayden's opinion, the O'odham were ancient inhabitants of the Sonoran Desert who were ruled by Hohokam overlords. Sometime in

the 1400s, other O'odham from the east, perhaps the San Pedro Valley, conquered the Hohokam and killed or chased away their chiefly shamans. He bases his scenario on a Gila Pima
(Akimel O'odham) creation narrative that he
himself recorded when he was excavating the
Hohokam site of Snaketown in 1935. That creation narrative vividly and hauntingly describes
the emergence of the O'odham from below the
ground and their destruction of Hohokam great
houses, beginning with Casa Grande (Bahr et al.
1994). Naviju and the wi:gida represent Pueblo
influences on the conquering O'odham when
they were living along the San Pedro and in close
contact with Pueblo peoples migrating southward from the Colorado Plateau.

The origins of the wi:gida—and the relationships among the O'odham, the Hohokam, and
late pre-Columbian Pueblo peoples—may never
be determined. But even if we never put the
entire mosaic together, the wi:gida remains a
fascinating piece of the Southwest's complex
culture history—as elusive as the smoke of the
campfires that warmed Hayden and his companions as they watched the Nanviju dancers half a
century ago.

onto fields. But whenever rainfall patterns varied or drainages shifted, the Tohono O'odham had to respond by following the water. As one farmer at the settlement of Ge Oidag west of Sells told Nabhan, "There used to be fields stretching on both forks. There used to be houses close to the edge of the fields, but then the wash changed and they had to move them" (Nabhan 1986b: 71). Aridity demanded mobility—an inescapable fact of life for the Desert People.

The Tohono O'odham Struggle for Water in the Twentieth Century

The pursuit of water changed somewhat during the twentieth century as the Tohono O'odham came under the jurisdiction of the United States government. About 18,000 members of the Tohono O'odham Nation occupy three reservations in southern Arizona. The largest encompasses 2,774,370 acres. South of the Gila and west of the Santa Cruz, it is a land of arroyos, not rivers. The other two—which are actually districts rather than autonomous jurisdictions of the Tohono O'odham Nation as a whole—straddle once-living rivers. The smallest lies along the Gila north of Gila Bend; the other surrounds Mission San Xavier del Bac. Located on the Santa Cruz south of Tucson, the San Xavier Reservation (Wa:k) is the wedge that has allowed the O'odham to force their way into the convoluted scramble for water in late-twentieth-century Arizona.

The scramble began more than a hundred years ago. Back then the Santa Cruz was what geographers call an intermittent stream. Much of its course was dry except during floods, but wherever geological formations forced its underground aquifer to the surface, spring-fed streams trickled through dense *bosques* (forests) of mesquite, cottonwood, and willow. One of those springs was Punta de Agua near San Xavier. For centuries, perhaps millennia, Archaic, Hohokam, and O'odham farmers irrigated their crops from its flow.

During the late 1800s, however, non-Indian farmers tried to intensify agriculture along the Santa Cruz by digging large ditches in the channel to intercept more of the water table. The strategy underestimated the terrible force of desert floods.

A Tohono O'odham adobe winter home in Fresnal Village in 1894. Winter villages were located in the foothills near permanent springs. The woman in the foreground is cooking corn. (Photographer unknown, Smithsonian Institution National Anthropological Archives.)

When heavy rains fell, runoff roared down those ditches instead of spreading across the floodplain as it had in the past. The ditches downcut, headcut, and ate away at the loose alluvial soil. By 1912 the Punta de Agua ditch had turned into a massive arroyo two miles long, six to twenty feet deep, and sixty to one hundred feet wide. The Sam Hughes ditch below Tumamoc Hill in Tucson was even more devastating. Between 1889 and 1912, it carved an eighteen-mile-long arroyo, carrying away 150 acres of prime O'odham farmland in the process. The water table dropped and fields were left high and dry. A living river was transformed into a barren chute that carried runoff away rather than slowing it down and allowing it to percolate into the desert soil.

That was the death of O'odham agriculture along the Santa Cruz. The federal government made a feeble attempt to keep the

A Tohono O'odham field and threshing floor with a summer field house near
Cold Fields, Arizona, in 1975. (Photograph by Helga Teiwes, Arizona State Museum)

corpse alive for a few years, but the Bureau of Indian Affairs
(BIA) "quit the pumps" in 1917 because they were too expensive.
Meanwhile, the city of Tucson was sinking more and more wells
into the Upper Santa Cruz Basin aquifer beneath the reservation. By the early 1960s, discharge was exceeding recharge by
nearly 100,000 acre-feet a year. In the words of one reporter for
the *Tucson Citizen* of February 6, 1980, the reservation's water
table had been "sucked so low that new wells hit dry bedrock."
The BIA had proved to be a miserable trustee of O'odham water
rights.

But the O'odham refused to dry up and blow away. In 1975
they pressured the federal government to file suit against the
city of Tucson, the corporate farmers, and the copper mines that
were plundering their aquifer. The legal foundation of the suit
was the Winters Doctrine, which arose out of the 1908 U.S.
Supreme Court decision *Winters* v. *United States*. The Winters

Doctrine proclaimed that Indians "have prior and paramount rights to all water resources which arise upon, border, traverse, or underlie a reservation." Moreover, that water had to be supplied in sufficient quantity to satisfy both the present and future needs of the reservation. The BIA had never tried to enforce the Winters Doctrine until the O'odham forced its hand. "More than a century of government failure to preserve the Papagos' interests assured the tribe of a court victory," an editorial in Tucson's *Arizona Daily Star* of June 14, 1982, observed, "and victory for the Papagos could have meant the permanent shutdown of mines and farms and an end to city growth and development."

The result was an endless series of negotiations to prevent Armageddon over water in the Tucson Basin. The first stage in that process culminated in 1982 with the passage of the Southern Arizona Water Rights Settlement Act (SAWRSA). That legislation granted the Tohono O'odham 76,000 acre-feet of water a year from three sources: 10,000 from groundwater pumping; 28,200 from Tucson effluent; and 37,800 from the Central Arizona Project (CAP)—the Holy Grail of Arizona water politics since the 1920s. That water had to be "suitable for agricultural use." Furthermore, the secretary of the interior was instructed to "improve and extend" irrigation systems on the San Xavier (Wa:k) district and the huge S-cuk Du'ag district on the main body of the reservation. The water could be used for any purpose, including lease to non-Indians, but the act also provided funds to "subjugate" reservation land and deliver CAP water to irrigate it. One possibility was the creation of a 10,000-acre commercial farm called the San Xavier Development Project. Another was the transfer of 20,000 acres of nonreservation farmland in the Avra Valley that had been purchased and "retired" by Tucson so that the city could use its water. Both possibilities presented the O'odham with an uncertain agricultural future that depended on massive investments of capital and continued federal intervention.

Ten years after the act was signed by President Ronald Reagan, however, it still had not been implemented. On the contrary, debate raged among the O'odham about what to do with the water when it arrived. O'odham leaders in Sells opposed the

DIET AND DIABETES AMONG THE O'ODHAM

Thomas E. Sheridan

During the 1930s, anthropologist Ruth Underhill recorded the autobiography of a ninety-year-old Tohono O'odham woman named Maria Encarnacion, whom everyone called Chona. Chona was born in the village of Kui Tatk (Mesquite Root) in the foothills of the Quijotoa Mountains and spent her childhood in a world governed by the rhythms of the Sonoran Desert. Young girls ran for miles each day to collect water in earthen jars. Young boys chased jackrabbits and cottontails and dug wood rats out of their nests. "Ah, how we could run, we Desert People," Chona recalled, "all the morning until the sun was high, without once stopping!" (Underhill 1936:7).

The Tohono O'odham planted corn, beans, squash, and other crops along the arroyos that flowed after the summer rains, but most of their food came from the desert itself. "All the year round we were watching where the wild things grew so we could pick them," Chona told Underhill. "Elder Brother planted those things for us. He told us where they are and how to cook them. You would not know if it had not been given. You would not know you could eat cactus stems and shake the seeds out of the weeds. Elder Brother did not tell the Whites that. To them he gave peaches and grapes and wheat, but to us he gave the wild seeds and the cactus. Those are the good foods" (Underhill 1936:10).

At the time Underhill took down Chona's words, however, the "good foods" were being replaced by beef, wheat flour, lard, and processed sugar. The government also drilled wells that allowed the Tohono O'odham to remain in their villages year-round. Running no longer was a cultural and economic imperative, and people whose bodies were adapted to cycles of feast and famine found themselves leading sedentary lives and eating foods high in sugar, cholesterol, and fats. By the early 1990s, more than half the Akimel O'odham over the age of thirty-five had adult-onset diabetes—the highest known rate of diabetes in the world—but the rate among the Tohono O'odham will soon eclipse that.

Many other desert-dwelling people, including Australian aborigines, have developed high rates of adult-onset (Type II) diabetes during the late twentieth century as well. A number of researchers in Arizona and Australia therefore decided to examine the traditional diets of these people to find out what protected them from diabetes in the past. They discovered that many desert plant foods—such as cholla buds, chia seeds, tepary beans, the pads of prickly pear cactus, and flour made from mesquite pods—kept blood sugars low and even. Composed of complex carbohydrates, these foods break down into simple sugars at a very slow rate. The process of digestion is slowed even further by the high quantities of gel-forming soluble fibers, or mucilages, found in many of these foods. When ingested, the gel forms a barrier between carbohydrates and the enzymes that dissolve them. Acting in concert, these "slow-release" carbohydrates and soluble fibers maintain the body's sensitivity to insulin and prevent blood sugars from increasing sharply after meals. Current estimates are that the O'odham of a century ago consumed four to five times the amount of soluble fibers their descendants do today.

Armed with these nutritional studies, groups like Native Seeds/SEARCH in Tucson are working with the Tohono O'odham and other tribes to promote a return to traditional foods, or at least a decision to incorporate more of these foods into their diet. The results have been promising. Earl Ray, an Akimel O'odham diabetic from the Salt River Reservation east of Phoenix, shifted from fast foods to tepary beans, cholla buds, mesquite flour, and chaparral tea. The five-foot-six-inch Ray dropped from 239 to 150 pounds and no longer required medication to control his diabetes. "We always kept gruel in our house," Chona told Underhill. "It was in a big clay pot that my mother had made. She ground up seeds into flour. Not wheat flour—we had no wheat. But all the wild seeds, the good pigweed and the wild grasses. . . . Oh, good that gruel was! I have never tasted anything like it. Wheat flour makes me sick! I think it has no strength. But when I am weak, when I am tired, my grandchildren make me gruel out of the wild seeds. That is food." We now know that Chona's memories were grounded on more than a romantic nostalgia for the past.

Akimel O'odham and Maricopas trade with Anglo American soldiers in 1846 during the Mexican War. (Lithograph by Major W. H. Emory)

Avra Valley option and advocated the San Xavier alternative. The San Xavier district council initially supported the farm, but then, in a sense, past generations of O'odham jumped into the controversy and changed the very vocabulary of the debate. In early October 1983, an enormous storm sent floodwaters surging down the drainages of southern Arizona. At San Xavier, those waters exposed a prehistoric burial site, triggering a chain of events that transformed the politics of reservation resource use—not just in the San Xavier district but across the country. Suddenly the disposition of American Indian human remains became a burning issue that dominated the lives of archaeologists, museum curators, and government bureaucrats, including those who were determining what to do with SAWRSA water.

As anthropologist Thomas McGuire points out, the issue was "not simply a procedural one" about what to do with bones and ash. On the contrary, it challenged the dominant ideology of resource use by addressing "a fundamental cosmology of the past." In a debate dominated by the "vocabulary . . . of law and economics"—one framed in terms of cost-benefit analyses and environmental impact statements—many O'odham argued that the well-being of the dead had to be assured as well (McGuire 1991:155). By 1989 the San Xavier district council had changed its mind about a commercial farm on reservation land. Juliann Ramon, a member of the council, said that if her ancestors were disturbed, her family would be "cursed with pain and disease until the seventh generation." Reservation land was living land that should not be scraped and subjugated. The past was not to be sacrificed to either the present or the future. As of 1995, the Tohono O'odham still had not decided what to do with their water.

Akimel O'odham Agriculture: From Abundance to Desolation

The Akimel O'odham were even more intertwined with the politics of water control in the American West. In the precontact period, the Akimel O'odham, unlike the Tohono O'odham, did not have to migrate between winter villages and summer camps. On the contrary, they lived in more-or-less permanent settlements along all the major rivers in the Sonoran Desert, including the Dolores, the Magdalena, the Altar, the Concepción, the Santa Cruz, the San Pedro, and the Gila. As a result, both their fields and their villages were larger than those of their Tohono O'odham brethren.

Nonetheless, the Akimel O'odham did face one major limitation: the lack of a major food crop that could be grown during the winter months, when frosts were a danger. Corn, beans, and squash—the staples of their diet—withered when the weather turned cold. In the upland areas, the River Pimas were able to cultivate their fields only from March through October, a constraint that restricted the size of their communities and their ability to depend on agriculture.

When Jesuit missionaries like Eusebio Francisco Kino began to visit the Akimel O'odham in the late 1600s, they brought gifts of Old World livestock like cattle and horses and Old World crops like winter wheat. Wheat, in fact, revolutionized Akimel O'odham society because it gave them a major cultigen that could be planted in November and harvested in June. By filling a gap in the agricultural cycle, wheat allowed the River Pimas to farm year-round. That in turn enabled them to live in larger, more permanent settlements, an important defensive measure as Apache raiding intensified in the eighteenth and nineteenth centuries. Wheat became so important to the Pimas along the Gila River that some even began to call the first month of the year Wheat Harvest Moon instead of Saguaro Harvest Moon (Ezell 1961; Russell 1975).

Wheat also turned the Gila Pimas into the first agricultural entrepreneurs in Arizona. During the war between Mexico and the United States, the Akimel O'odham traded both corn and wheat flour to the U. S. Army. After the war ended and gold was discovered in California, thousands of "argonauts" trudged through Pima villages along the Santa Cruz and Gila on their way to the goldfields, stimulating the Pimas to intensify production and expand their acreage. The market continued to grow with the establishment of a stagecoach route through Pima territory in 1857 and the outbreak of the Civil War in 1861. By 1870, Akimel O'odham along the Gila River were selling or trading several million pounds of wheat a year. Pima fields had become the breadbasket of Arizona Territory. In recognition of their importance, the federal government created the first Indian reservation in Arizona for the Akimel O'odham and their Yuman-speaking Maricopa neighbors in 1859. Known as the Gila River Indian Community, it was enlarged seven times between 1876 and 1915 until it encompassed 371,929 acres.

But the crucial resource in central Arizona was water, not land. It is interesting to speculate about what might have happened if Pima water rights had been respected. O'odham canals might have rivaled those of the prehistoric Hohokam. The Pimas might have diversified into milling, freighting, and stock feeding as well. But that was not to be. Beginning in the late 1860s, Anglo farmers began settling upriver along the Gila in the vicinity of Florence and Safford, digging canals and diverting the water of the Pimas. Crops failed and fruit trees withered.

In 1873, Antonio Azul led a delegation of Akimel O'odham to Washington, D.C., to protest the situation. The government re-

Men with horse teams building the Florence canal in 1887. (Arizona Historical Society)

sponded by suggesting that the Pimas move to Indian Territory in Oklahoma. The Pimas refused, but 1,200 did migrate north to the Salt River, where irrigation water was available. The Salt River Indian Reservation was set aside for those migrants, including several communities of Maricopas, in 1879. The rest of the Akimel O'odham fanned out along the Gila and farmed as best they could along stretches of the floodplain where water still flowed.

Then in 1887 the construction of a large canal near Florence diverted the Gila once and for all. By 1895 conditions were so desperate that the government had to issue rations to the Pimas. Pima calendar sticks, which had recorded floods and Apache attacks three decades before, now said, "There was no crop this year," or, "This year there was no water." In the span of a single generation, the Akimel O'odham had been transformed from prosperous independent farmers into poverty-stricken wards of the state. Once they had produced enough food to feed the territory; by the turn of the century they did not even have enough water to grow food for themselves.

The Akimel O'odham Struggle for
Water in the Twentieth Century

During the first half of the twentieth century, the federal government made some half-hearted and contradictory attempts to restore water to the Akimel O'odham. Its most ambitious effort was the San Carlos Project, which was approved by Congress in 1924. The government was not about to take water away from powerful non-Indian farmers in Florence and Safford, so it decided to wrest more water from the Gila River by building an enormous dam. The result was Coolidge Dam, a strange, bulging structure designed to create a reservoir capable of holding 1.3 million acre-feet of water. Its capacity was determined by an estimated average annual runoff of 460,000 acre-feet—an estimate developed during the abnormally wet period of 1899 to 1928. Unfortunately for the O'odham, however, six decades of man-made drought caused by upstream farmers were followed by a long-term regional drying trend that reduced runoff to an average of 215,000 acre-feet a year. Looking out at the huge, shallow, vegetation-choked pond during the dam's dedication in 1930, Will Rogers quipped, "If this were my lake, I'd mow it." San Carlos Lake never filled to capacity until 1979, fifty years after its completion. The rest of the time it was usually less than two-thirds full—an enormous hydrological blunder that mocked O'odham dreams of an agricultural future.

But if Coolidge Dam never impounded the water it was supposed to impound, it did control floods. And without floods to replenish alluvial aquifers on the Gila reservation, springs dried up, trees died, and at least twenty-nine species of birds disappeared from the Gila's banks (Rea 1983). The last few oases of Pima agriculture turned to barren landscapes—bleached salt flats, skeletal mesquite bosques, empty irrigation canals. Upstream agriculture and Coolidge Dam killed the middle Gila River. When it died, the last vestiges of Pima self-sufficiency died with it. George Webb, a Pima, captured the desolate poignancy of that demise when he wrote:

> In the old days, on hot summer nights, a low mist would
> spread over the river and the sloughs. Then the sun would

Coolidge Dam along the Gila River. (Arizona State Museum)

come up and the mist would disappear. On those hot nights the cattle often gathered along the river up to their knees in the cool mud.

Soon some Pima boy would come along and dive into the big ditch and swim for awhile. Then he would get out and open the headgate and the water would come splashing into the laterals and flow out along the ditches. By this time all the Pimas were out in the fields with their shovels. They would fan out and lead the water to the alfalfa, along the corn rows, and over to the melons. The red-winged blackbirds would sing in the trees and fly down to look for bugs along the ditches. Their song always means that there is water close by as they will not sing if there is not water splashing somewhere.

The green of those Pima fields spread along the river for many miles in the old days when there was plenty of water.

Now the river is an empty bed full of sand.

Now you can stand in the same place and see the wind tearing pieces of bark off the cottonwood trees along the dry ditches.

The dead trees stand there like white bones. The red-wing

blackbirds have gone somewhere else. Mesquite and brush and tumbleweeds have begun to turn those Pima fields back into desert. (Webb 1959:124–25)

The federal government could not, or would not, revive the dead river, but it did try to keep Pima agriculture alive. Nonetheless, its efforts were plagued by paternalism, mismanagement, and the rapid commercialization of Arizona agriculture. The amount of land under cultivation on the Gila reservation rose from 12,000 to 30,000 acres during the 1930s, but most of that land was leased to non-Indian farmers and was irrigated by pump-powered wells. There was no more water in the river to farm the old way, and the new way required new skills and massive investments to pay for insecticides, chemical fertilizers, machinery, and fossil fuels. That meant a reliance on federal funds and federal supervision. The Pimas who tried to play by the government's rules found themselves being told when to irrigate and what crops to raise. "Under the San Carlos Project," anthropologist Robert Hackenberg concluded, "the Pima were farm owners, but they were no longer operators" (1983:175).

By 1937 the Akimel O'odham were so disgusted by the project's empty promises and bureaucratic red tape that their newly formed tribal council refused to pay for its operation or maintenance. By then, about the only O'odham left in the agricultural sector were thousands of farm laborers who picked cotton every fall in other people's fields. Even those jobs disappeared during the 1950s, when cotton harvesting became mechanized. A survey in 1961 uncovered only ninety-seven individuals on the Gila and Salt River reservations who listed their occupation as farm owner-operator. Twenty years later, 103 members (8.8 percent) of the Gila reservation's Indian workforce of 1,160 people toiled on tribal farms. More than twice that number (208) were employed by the three industrial parks on reservation land, while 44 percent (508) worked for the tribe itself. Most of the rest had jobs with the Bureau of Indian Affairs, the Federal Indian Health Service, or the public schools.

Today the Akimel O'odham continue to press their claims to Gila River water through a maze of complex and expensive lawsuits. For most of the Pimas, however, farming is little more than an old dream, a collective memory of a time when their river flowed and birds sang in green and living trees. Nonetheless, the dream refuses to die. Some Pimas see their future in education and industrialization, but others, like Lester Lewis, continue to equate "lasting well-being" (ká·cim ʔapʔedag) with life on the land. Speaking to a meeting of the conservative Gila River Farmers Association, Lewis said:

> When I was on the tribal council there were requests to lease land. I opposed these requests because we are a growing tribe and we are crowding up this ground that is our property. The land is ours and we own a portion of that water and will use it to plant crops. The ones who want to lease our lands are saying that the O'odham are lazy and are not farming. They say that we don't know how to farm. It is not so. (Pablo 1983:215)

Conclusion

For both the Akimel and Tohono O'odham, then, the future remains clouded by water issues—how to get it back and what to do with it if and when it comes. More and more, however, the O'odham are calling for that water in their own cultural idiom, not just in the language of lawyers, economists, and hydrologists. Whether the government or the other parties in the struggle will listen remains to be seen.

Nevertheless, the old songs continue to be sung. Many O'odham move off the reservations, but others return with a renewed vision of what it means to be O'odham in the modern world. Bernard Fontana wrote of one such individual, Danny Lopez, who returned to found the Desert Indian Dancers after attending a Franciscan boarding school and working in a copper mine south of Tucson:

> Together, we visited Gu Oidak. Before this, we had stood together on the summit of Baboquivari Peak, the physical and spiritual center of the Papago universe. So had we stood on the roof of Mission San Ignacio, on a hundreds-of-years-old trail west of Sil Nakya, and at the edge of the pond at Quito-

vac. Now, however, we were home. Not my home, but Danny's home.

Growing near the house where his parents live in Gu Oidak is a mesquite tree. He pointed to a spot on the ground beneath the tree. "That," he said, "is where I was born."

I marvelled. How many of us, I thought to myself, can point to a place on the earth and say with honesty it marks the location where we first emerged into sunlight or starlight? Here was where Danny first breathed the desert air; this is where he uttered his first cries as a human being. Not twenty miles away "somewhere," not even a mile away nor ten feet to the left or right. But here. On this very spot!

Emergence for most of us is a hospital happening, an event surrounded by medical paraphernalia and the studied indifference of our medical subculture. Hospitals, like their personnel, come and go. Small wonder that we lack a sense of place, of roots. We are born in Rochester or Phoenix or Denver or New Orleans. But Danny was not born in Gu Oidak. He was born *there*, under that mesquite tree on that piece of ground. Roots and a sense of place? How could it be otherwise? (Fontana et al. 1981:143–44)

THE RARÁMURI (TARAHUMARAS)

WHEN WE WALK IN CIRCLES

Thomas E. Sheridan

In 1887 a Norwegian explorer named Carl Lumholtz decided to find out if there were any "Cave-Dwellers" left in North America. Four years later, after crisscrossing the continent to raise money for the expedition, Lumholtz and his party ventured into the rugged Sierra Madre Occidental of northwestern Mexico, where they encountered the Tarahumara Indians. "I determined to study these interesting people, especially the so-called *gentiles* (pagans)," wrote Lumholtz in his two-volume classic, *Unknown Mexico,* "and as this was not practical, even with the present reduced size of the expedition, I gradually disbanded the entire company and at last remained alone" (Lumholtz 1902:xii).

The explorer lived with the Tarahumaras for the next three years, recording his observations of their religious beliefs and material culture. According to Lumholtz, non-Christian Tarahumaras had two gods, Father Sun (Nonorúgame) and Mother Moon (Yerúgame), whose son was Morning Star. And while Christian Tarahumaras may have adopted Catholic images of God the Father (Tata Dios), the Virgin Mary, and Jesus Christ, these foreign religious figures had been transmuted into the older Native American trinity. Christian Tarahumaras also added the Devil to their pantheon. As Lumholtz wryly noted, "The Tarahumares are the sons of God, and the Mexicans the sons of the Devil. For this reason the Tarahumares say that it is no crime to eat the cows of the Mexicans; they think the cows do not really belong to the Shabotshi [*chabochi,* non-Indians] anyway" (Lumholtz 1902:296).

Lumholtz went on to relate several Tarahumara myths about the creation of the universe, including one entitled "The Sun and the Moon in the Beginning of the World":

> In the beginning the Sun and the Moon were alone, and they were children. They wore dresses made of palm-leaves, and they lived in a house thatched with palm-leaves. They had neither cattle nor sheep. Both the Sun and the Moon were dark, and the Morning Star was the only one that shed any light on the earth. The Moon was eating lice from the hair of the Sun, and the Morning Star was watching at night. There were 600 Tarahumares at that time, and they were much hampered by the darkness. They could not do their work, and they had to hold each other's hands, and they were stumbling all the time. Then they cured the Sun and the Moon by dipping small crosses into tesvino (corn beer), and touching the Sun and the Moon on the chest, on the head, and on the back. Then the Sun and the Moon began to shine and to shed light. (Lumholtz 1902:297)

Later anthropologists, especially William Merrill of the Smithsonian Institution, have described Tarahumara cosmology with greater sophistication. Nevertheless, two of Lumholtz's observations ring as true now as they did then. First, the Tarahumaras did not assimilate Christianity; they made it their own. They may have incorporated some of the trappings of Roman Catholicism into their culture, but their world view remained distinctly Native American. Second, even though they respected the power of God and the Devil, the Tarahumaras saw themselves as actors, not reactors, in the cosmological drama. They cured the Sun and Moon when the world was young just as they protect God from the Devil every Easter. Such cultural tenacity is one of the reasons why the Tarahumaras have survived the intrusions of outsiders—Spaniards and Mexicans, miners and missionaries—for more than four hundred years.

The Land and the People

The Tarahumaras call themselves Rarámuri, a term that means human as opposed to non-human and Indian in contrast to non-

Indian. About 60,000 Tarahumaras live in the state of Chihuahua, and more than 1,000 in Sonora. That makes the Rarámuri
the second-largest American Indian group in the Greater Southwest, exceeded only by the Navajos. The Tarahumaras speak a
Uto-Aztecan language related to both Piman and Cáhitan
(Yaqui-Mayo). Fragmentary archaeological evidence suggests
that ancestors of the Tarahumaras may have occupied the valleys and canyons of the Sierra Madre for at least the past 2,000
years.

Geography dominates Rarámuri society and history. In the
late sixteenth century, when Spaniards first snatched at the edges
of Tarahumara territory, the Rarámuri occupied more than 5,000
square miles in southwestern Chihuahua. It is an extraordinarily
wild and rugged land. Eastern Tarahumara country lapped over
onto the high Chihuahuan plains, but the heart was the northern
Sierra Madre Occidental, Mexico's "Mother Mountains," the
long, curved backbone of the nation that sweeps from the U.S.
border to Jalisco. Approached from the east, the Sierra seems
unspectacular—a rolling plateau that gradually rises to rounded,
pine-clad peaks 8,000 to 10,000 feet high. But if you head south
and west, crossing the continental divide, you plunge into the
barranca (canyon) country—a jagged labyrinth of gorges cut by
rivers that tumble from the Sierra to the Pacific. Some of those
gorges, such as the famous Barranca del Cobre of the Río Urique,
are more than a mile deep in places. By climbing up and down the
steep slopes, the Rarámuri were able to move from the forest to
the tropics in less than a day, collecting pinyon nuts, acorns,
cactus fruit, and agave hearts as they went.

To the east, where they farmed the floodplains of the Río
Conchos and its tributaries, the Tarahumaras may have lived a
more sedentary life based on irrigation agriculture. In the Sierra
Madre, however, the rough terrain compelled the Rarámuri to
rely more on hunting and gathering and the cultivation of corn,
beans, and squash. Small groups of related families inhabited the
mountain valleys, planting their crops with digging sticks along
streams. They also shifted their residence to cultivate fields in
different areas and to take advantage of wild resources. This
transhumance (seasonal movement) was particularly pronounced in the canyon country, where dramatic changes in ele-

THE RARÁMURI AND THE LEADVILLE TRAIL 100

Thomas E. Sheridan

Most Native peoples of the Greater Southwest ran great distances on a daily basis. Many groups also played local variants of a kickball game that the Opatas called *gomi* and the Tarahumaras *rarahipa*. Many scholars believe that the word *rarámuri* itself means footrunner.

In Cucurpe, Sonora, the last gomi was held in 1949. Among the Rarámuri, however, the kickball game continues to be played at *tesgüinadas* (drinking parties), death rituals (*napisós*), and special events where teams from two or more communities race against one another. The persistence of the rarahipa is one of the reasons why the Rarámuri have acquired legendary reputations as long-distance runners. Stories abound of Tarahumaras racing twenty-one miles in two hours or covering five hundred miles in five days. During the early 1960s, physical anthropologists, physiologists, and physicians even studied Rarámuri runners and found that their resting heart rate and blood pressure were 20 percent below the norm. "Probably not since the days of the ancient Spartans has a people achieved such a high state of physical conditioning," one doctor exclaimed (Fontana 1979:106).

But road building and other encroachments of modern life were threatening the Tarahumaras' reliance on running, so in 1992 *chokéame* (race judge) Patrocinio López organized a team of five Rarámuri runners to participate in the Leadville Trail 100, one of the most grueling endurance races in the world. "Ultrarunners" from across the globe come to Leadville, Colorado, to run the 100-mile course, which begins at an elevation of 10,000 feet and twice climbs 12,600-foot Hope Pass. López felt that, by competing, the Tarahumaras could reinforce their commitment to running as a way of life.

The results of the 1992 race were disappointing. None of the Rarámuri completed even half the course. The next year, however, two members of the 1992 team and four new runners returned to Leadville. Having analyzed their mistakes, they also traded in their new running shoes for huarache sandals made from tire treads scavenged from the Leadville dump.

The race began at 4 A.M. and the Rarámuri started conservatively. Then they reached Hope Pass. The trail was steep and rocky, like the trails in the Barranca del Cobre back in the Sierra Madre. By 72 miles into the race, 30-year-old Manuel Luna had moved into second place, but 38-year-old Cerrildo Chacarito and "38-year-old" Victoriano Churro were closing quickly, nourishing themselves with sips of pinole, a drink made of cornmeal mixed with water. Victoriano crossed the finish line in first place with a time of 20 hours, 2 minutes, and 33 seconds. Cerrildo was second with a time of 20:43:06, and Manuel fifth at 21:26:09. In a race where more than half the runners drop out, only one member of the Tarahumara team failed to finish. It was a stunning vindication of the Rarámuri legend, and the crowd went wild.

That legend only grew stronger when race officials discovered that the winner, Victoriano, was in fact a 55-year-old grandfather, the oldest winner of the race. U.S. immigration officials had asked Victoriano his year of birth. Misunderstanding the question, Victoriano responded with his age. In the words of ultrarunner Jamie Williams, who paced Victoriano, "It was beautiful—running at night, in the lightning, the snow, the rain, next to the streams, down narrow paths—running with a man who ran like a deer" (Williams 1994:27).

vation, temperature, and vegetation provided access to a wide variety of plant and animal foods.

The Tarahumaras practiced slash-and-burn agriculture in the barrancas as well, a method of cultivation they called *mawechi* and the Spaniards dubbed *rosa*, from the verb *rozar*, to clear a field of vegetation for planting. Franciscan missionary Joseph Agustín Falcón Mariano, who wrote a detailed *relación* of the mission of Guaguachique in 1777, described how the rosa system worked in the Barranca del Cobre:

> In February and March these slopes are cleared of brush and brambles, which are left until May or early June, when the rubble is burned. They take care to ensure that all the waste burns and that the soil is evenly coated with ashes. These fields are then planted with the first rains, using digging sticks. . . . When the soil is good a cleared field is used for two or three years. After three years have passed these fields no longer produce, and must be fallowed until wild growth returns to cover them. (Sheridan and Naylor 1979:109–10)

Thousands of Tarahumaras lived in this immense land, ranging from the pine-clad mesas of the Río Papigochic drainage in the north to the tropical canyons of the Río Verde in the south. But the Rarámuri were never a unified tribe with centralized political authority or compact towns. On the contrary, they scattered themselves across the landscape in dispersed settlements the Spaniards called *ranchos*, each of which was more or less independent of the others. Headmen and councils of respected elders provided leadership, but decision making was a diffuse process that depended more on consensus and personal charisma than coercion. Unlike the Europeans, the Tarahumaras did not devise a hierarchical society dominated by kings and priests. That gave them great flexibility, but it also limited their ability to mobilize against the European threat.

Because very little archaeological research has been done in the northern Sierra, it is hard to discern even the outlines of aboriginal Tarahumara culture. Nevertheless, documents written by early Jesuit missionaries provide tantalizing glimpses of that culture at the beginning of the seventeenth century. In many respects the society described in those fragile handwritten

Rarámuri in front of their thatched houses and stone storehouses in the early 1890s. (Photograph by Carl Lumholtz, American Museum of Natural History)

pages differed little from the one Lumholtz found nearly three hundred years later. Consider, for example, the words of Padre Juan Fonte, a Catalan from Barcelona who visited the southeastern Tarahumara country sometime between 1608 and 1611:

> Many of these Indians live in caves, of which there are many. Some are so large that extended families can live in them, dividing the interior into smaller dwellings. They dress themselves with *pita* cloth [usually made from the fibers of agave]. The women are expert in weaving this cloth and dress themselves in it as well. The women are very shy, and neither sit nor mix with the men. The Tarahumara burial customs are different from those of other Indian nations, in that, the dead are buried in a separate and designated place, much like a cemetery. Personal belongings and food for the spiritual trip are placed with bodies. The house in which the person died is either burned or totally abandoned, and the relatives of the dead mourn by cutting their hair. (Sheridan and Naylor 1979:9)

Warfare was apparently common in the early seventeenth century, not only between the Tarahumaras and their southern neighbors, the Tepehuanes, but among the Tarahumaras themselves. In 1638, Padre Gaspar de Contreras, superior of the new missions of the Tarahumara Baja in the eastern foothills, noted "their fear of war, which is always with them, because this same nation often engages in fierce battles among themselves in which a great many people die miserably" (Sheridan and Naylor 1979:11). Such battles undoubtedly took place before the arrival of the Europeans—for land, for water, for masculine prestige. But the advance of the Spanish empire aggravated those conflicts by forcing groups into the territories of others and by creating a growing market for slaves. Slave raids—and the slave trade—linked American Indians from central Mexico to the Great Basin. No group, not even the Rarámuri in their mountain heartland, was immune.

Miners and Missionaries

Chronic warfare was one reason many Tarahumaras joined the mission communities that Padre Fonte established along the Río Balleza, a tributary of the Conchos. Many Indians, including the Yaquis on the other side of the Sierra, saw the early priests as peacemakers—powerful black-robed shamans with metal tools and good things to eat who were able to impose an imperial Spanish peace on groups like the Tarahumaras and the Tepehuanes. Fonte and his colleagues convinced hundreds of members of both groups to settle next to one another under the protection of Both Majesties—the Catholic King and the Catholic God.

But peace did not last long. In 1616 the abuses of Spanish miners and the demands of European missionaries drove the Tepehuanes to launch one of the bloodiest nativist rebellions in the history of the Greater Southwest. The uprising was led by a religious visionary named Quautlatas, who called for the Indians to drive out the Spaniards and return to their old gods. He declared that all Indians killed in the revolt would be resurrected and that no Spanish reinforcements would arrive because their ships would be sunk. Fueled by such beliefs, the rebellion raged for two years, killing more than four hundred Spaniards and six of the seven Jesuits assigned to the Tepehuan and Tarahumara missions. Fonte was one of the victims. The conversion of the Tarahumaras did not resume until the late 1630s, when Jesuits founded the mission province of Tarahumara Baja, or Lower Tarahumara, in the Río Conchos drainage.

By then, miners had overwhelmed missionaries as agents of exploitation and change. When Spaniards first settled east of the Rarámuri in the 1560s, the Tarahumaras traded corn for cloth and other European goods. That pattern may not have been much different from late prehistoric patterns, when the great center of Paquimé (Casas Grandes) in northwestern Chihuahua dominated trade between Mesoamerica and the Southwest. Shells and Casas Grandes pottery have been recovered from at least one site in Tarahumara territory. The Rarámuri may have been exchanging corn, hides, and other resources for the exotic items fashioned by the craftsmen of Paquimé, who turned the raw materials of half a continent—turquoise, feathers, sea shells—into the stuff of power, prestige, and dreams.

Beginning in the 1630s, however, the pattern shifted. Massive silver strikes north of Santa Bárbara attracted thousands of miners, who founded San José de Parral. Merchants and ranchers followed, and Parral soon became the most important settlement in the region. Rarámuri continued to trade with the newcomers. Some even descended into the mining communities to work for wages in the dark shafts and mercury-poisoned reduction works, where many of them died. But thousands more were rounded up in labor gangs under the *repartimiento* system, which distributed them to Spaniards to dig ore, make charcoal, and harvest fields. A rotational labor draft, the repartimiento system was only supposed to take no more than 2 to 4 percent of a community's able-bodied males at a time. It also operated under numerous other regulations designed to prevent the abuse of Indian labor. But as more Spaniards poured into the area—and more Indians died of smallpox, measles, and other Old World diseases—the regulations were ignored and abuses proliferated. Some Indian communities lost up to a third of their workforce at a time.

Faced with such mistreatment, it was only a matter of time
before many Rarámuri rebelled. A wave of uprisings broke out
in 1648, 1650, and 1652. The revolts were crushed with the help
of Tarahumara allies like Don Pablo, a former rebel leader who
tracked down four of his fellow chieftains and presented their
heads to the Spaniards. An even bloodier cycle of insurrections
swept across Tarahumara country four decades later. In 1687,
silver was discovered at Cusihuiriachic in northern Tarahumara
country. More veins of silver were found, more miners descended upon the region, and more Tarahumaras were impressed as woodcutters, house builders, and burden bearers.

At the same time, epidemics of smallpox and measles devastated the Indians in 1693 and 1695. "Because of the plague,"
observed Padre Joseph Neumann, "some of the magicians
sought to persuade their Indians secretly to withdraw from the
villages and go farther away from the sound of the church bells
which, they said, attracted diseases." Neumann, who firmly believed in the demonic power of the shamans, went on to note,
"They affirmed, moreover, that baptism was a source of con-
tagion to children; that in reality the missionaries were the wiz-
ards and sorcerers of the Spaniards, and that the Indians should
be on their guard against them" (Sheridan and Naylor 1979:46).

The result was rebellion, which flared briefly in 1690 and then
exploded in 1697. Juan Fernández de Rentana, the Spanish com-
mander, retaliated with a war of blood and fire, executing thirty
of the rebel leaders. "As a salutary example to the rest," wrote
Neumann, "their heads were cut off, set up on spears, and displayed to the public view; some of them in Cocomórachic, others
along the road to Yepómera. Thus they were left, to become the
food of crows" (Sheridan and Naylor 1979:50).

The uprising of 1697–98 was the largest and last major Tarahumara revolt. But even though they never again took up arms
against the Spaniards in significant numbers, they had other
ways of resisting. Tarahumara *principales* (headmen) gave *tlato-
les* (public speeches) railing against the evils of baptism. When
missionaries cornered individual Indians, many said they were
too old to be baptized or had already been baptized, or they
engaged in long, convoluted rebuttals of Christian doctrine. The

Rarámuri also used sarcasm to denigrate the priests. "A donde vienes, Padre Cornudo? (Where are you going, Father Cuckold?)" they asked one missionary. European authority could be challenged with humor as well as bloodshed.

Retreat and Reducción

Above all, however, the Rarámuri retreated—deeper into the mountains, deeper into the barrancas. The introduction of Old World animals, especially sheep and goats, reinforced the old transhumant way of life, allowing the Tarahumaras to intensify their use of land that could not be cultivated. Manure from those animals also enabled them to intensify their agriculture. Rarámuri began penning their herds in movable wooden corrals that they placed in their fields each night, renewing the fertility of the soil section by section. Oxen and metal plows gave them the technology to expand their acreage. Wheat, oats, peach trees, and vegetables increased their agricultural repertoire. Not everything European was rejected.

Most missionaries did not appreciate the ecological soundness of seasonal movement or dispersed settlements. One of the foundations of the mission program, in fact, was the policy of reducción—the process of gathering the Indians from the wilderness and "reducing" them to mission life. Most missionaries came from Mediterranean Europe, where cities were equated with civilization. They saw people who lived outside the cities—gypsies, shepherds, Indians—as savages. Missionaries viewed wilderness as a source of disorder and a refuge of hechiceros (witch doctors, the Spanish term for Native American shamans), who communed with the Devil and turned their backs on God, who could only be reached through the intercession of a priestly hierarchy embodied by the Roman Catholic Church.

No one expressed this cultural chasm better than the acerbic Padre Juan Isidro Fernández de Abee, who worked among the Rarámuri in the 1740s. According to Abee,

> The Indians live at great distances, separated from each other and scattered. They prefer to live, and have their dwellings, in the ravines and canyons of the cold, inhospitable mountains.

Their native simplicity, barbarity, laziness, and sloth incline them to this in preference to being reduced to living rational, civilized human life. . . . From infancy they are brought up to be mountain vagabonds with no training in either morals or proper conduct. They are worse than beasts, because the animals at least acknowledge their subjection to those who put them in cages and govern them. (Sheridan and Naylor 1979: 79, 81)

Abee also had no faith in friendly persuasion. In his words,

Even when the zealous ministers collect them into pueblos by repeated effort, vigilance, discomfort, and dangerous travel, the Indians are neither stable nor do they stay in the pueblos. This is why, for the honor and glory of God and the service of the King our Lord, the royal ministers should cooperate with the armed forces and give aid to the tireless laborers in the vineyard by repeated expeditions into the sierra. They should remove the Indians from their canyons and use armed force to make them live together in pueblos. They should burn their hovels and *sonogoris* [corncribs] in which they keep their corn. Eight or ten soldiers ought to be stationed at this mission until the Indians learn to love their pueblo. If they flee from the pueblo after being settled there the soldiers ought to hunt them and bring them back. In this way the will of Our Catholic Kings will be done and some useful purpose will be served by the sweat poured out by the padre missionaries for the salvation of the wretched Indians' souls. (Sheridan and Naylor 1979:85)

Abee's attitudes reveal a fundamental fact about the mission program. No matter how humane individual missionaries may have been—and there were many who were true champions of the Native Americans—the mission endeavor was almost always backed by the threat of force. The padres and their Spanish patrons were determined to change the way the Indians worshipped and the way they lived. Thousands of American Indians voluntarily accepted baptism because they were attracted to the new religion or because they feared disease and thought that the Christian rituals would protect or cure them. But once they were converted, they were not allowed to renounce their faith. That was apostasy, which was considered as bad as heresy. In the seventeenth and eighteenth centuries, religious freedom did not flourish in the Spanish empire or any other European domain.

The Tesgüinadas

Another custom the missionaries objected to was the *tesgüinada*, or drinking party. *Tesgüino*, which the Rarámuri call *suguí* or *batári*, is a fermented beverage with an alcoholic content similar to beer. It is usually made out of sprouted corn, although the baked hearts of agave or sotol are sometimes used as well. Tesgüino permeates Tarahumara culture. The Rarámuri consider it a sacred gift from God (Onorúame). They place *ollas* (pots) of it at the foot of crosses in curing ceremonies and during the rituals of Holy Week. Shamans (*owerúames*) sprinkle it on animals and corn to protect them from lightning and to ensure their growth. They also rub it on the lips of sick children and then make them drink some. Lumholtz even believed that without tesgüino, the Tarahumaras could not reproduce themselves. Tarahumara men, he said, were so shy they needed tesgüino to work up enough courage to have sex with their wives.

Later scholars have punctured that notion, but they do attest to the enduring importance of Rarámuri drinking parties. According to anthropologist John Kennedy, who lived among non-Christian Tarahumaras (*gentiles*) in the early 1960s,

> To the gentiles of Inápuchi, the tesgüinada serves all the functions of social life outside those served by the household groups. The tesgüinada is the religious group, the economic group, the entertainment group, and the group at which disputes are settled, marriages arranged, and deals completed. It is at the tesgüinada with his set of neighbors, and under the influence of alcohol, that the individual has a chance to play the roles in his repertoire which remain dormant during his isolation. (Kennedy 1978:117)

Kennedy argues that the "tesgüino network"—the group of people who generally exchange labor and drink together—is "the major structural form of the Tarahumara social system above the

Tarahumara *tesgüino* paraphernalia. Clockwise, from top center: a tesgüino jar
made by Refugio Chancarito of Panalachic in 1967; a spoon made before 1977; a
bowl made by the potter Rorana in Choguita before 1977; a gourd bowl made
before 1977; and a basketry strainer made in Guacaibo before 1977. (Photograph
by Helga Teiwes, Arizona State Museum)

family and residence group" (1978:125–26). He estimates that
most Tarahumara adults attend forty to sixty drinking parties a
year, devoting at least a hundred days to the tesgüinadas. Drinking parties bring large numbers of people together on a regular
basis to carry out all the tasks individual families cannot accomplish on their own: clearing land, building dwellings and storehouses, holding curing ceremonies for fields and animals as well
as humans. But it is also the major form of psychological release
in Tarahumara culture—the bond that links the scattered families
of the Sierra together and allows them to escape their isolation in
an explosion of drunken camaraderie. "When one approaches a
household where a tesgüinada is going on, he is immediately

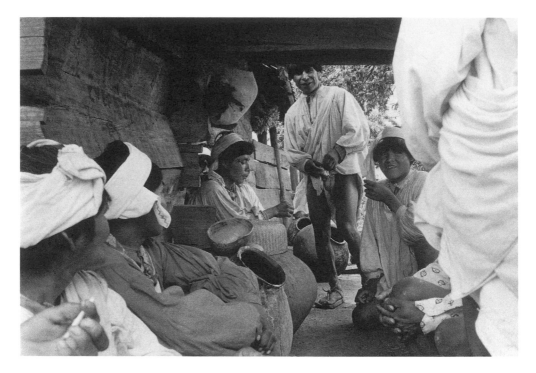

A *tesgüinada* at Rejogochi in 1981. (Arizona State Museum)

impressed by the din," Kennedy observed. "The loud hum, punctuated by shouts and singing, mixed with noise of drunkenly
operated violins, guitars, accordions, and harmonicas, is a shock-
ing contrast to the usual peaceful silence of the rancho, broken
only by raucous barks of mongrels" (Kennedy 1978:117).

Not surprisingly, the missionaries viewed such gatherings
with alarm. To them the tesgüinadas were nothing but drunken
orgies where the Devil reigned supreme. When Padres Tomás de
Guadalajara and José Tarda visited the Tarahumara community
of Tutuaca in 1675, the two Jesuits scurried up the rocks of the
highest mountain as soon as they saw the Indians bring out their
"ollas of wine." The next morning the Tarahumaras and Pima
Bajos living there asked why the priests had not baptized them.
The missionaries retorted:

> We told them that we had not come to applaud their drunken
> debauchery but to eradicate it. We also said that it was not
> decent for persons of our state to find ourselves in a house
> where God, whose priests we were, was offended, and that we

would not baptize drunkards. With this one astonished man asked, "Are these feasts of sin?" The man did not know that they were and the others agreed. They then said that because the Christian Spaniards did not permit such things, they would forsake them at once and never become drunk again. (Sheridan and Naylor 1979:35)

When We Walk in Circles: Holy Week among the Tarahumaras

The missionaries may have won a temporary victory at Tutuaca, but the tesgüinadas continued. Drinking parties were too important to Tarahumara society. They knit the Rarámuri together, mobilizing labor and relieving the isolation of life in a rugged land.

Nonetheless, the Jesuits and their Franciscan successors did transform many aspects of Tarahumara life. One of the most enduring changes was political. The missionaries were rarely able to congregate the Tarahumaras in compact villages called pueblos. The thin soils and broken topography of the Sierra simply did not provide many locations where large numbers of people could live together—and farm together—year-round. But if the priests and Spanish officials could not create actual pueblos, they could thrust pueblo political organization upon neighboring ranchos, which is exactly what they did. The missionaries erected mission churches throughout both the Tarahumara Baja and Tarahumara Alta. The churches became the administrative as well as religious centers of the surrounding region. They were where the baptized Rarámuri (*bautizados*) of a particular "pueblo" gathered to receive religious instruction and to participate in Catholic rituals, and they were where the adult males of the region met to elect the *gobernador* (governor) of the pueblo and his assistants, who were also known by Spanish terms (such as *capitán, teniente,* and *fiscal*). The pueblo, with its formal hierarchy of officials, was a streamlined model of Spanish municipal government imposed on the Tarahumaras from above.

In characteristic Rarámuri fashion, however, the Tarahumaras adopted the institution and made it their own. Geographer Campbell Pennington estimates that the average pueblo encompasses an area with a radius of about fifteen miles—a territorial unit considerably larger than the autonomous ranchos of the past (Pennington 1983). That means that more Tarahumaras come together on a regular basis than they did before missionization began. Nonetheless, the ancient rancho settlement pattern still survives. The Tarahumaras accepted pueblos, but they rejected *reducción*.

They utilize the offices of the pueblo for their purposes as well. The governor, called *síriame* in Rarámuri, and his assistants organize religious fiestas, resolve disputes, and punish wrongdoers. They also deliver the sermons that have become such an important part of Rarámuri culture. According to anthropologist William Merrill, "the primary purpose of such speeches is not to communicate new information but to reiterate time-proven advice for the proper conduct of life" (1988:62). The advice transmits values and beliefs that are far older than the Catholic teachings of the seventeenth and eighteenth centuries. Delivered in Rarámuri rather than Spanish, the sermons of a governor to his people usually begin by invoking the two most important Rarámuri deities—Our Father (Onorúame) and Our Mother (Iyerúame). The governor then urges the people to ask their forgiveness and to follow their paths. But forgiveness in the Rarámuri moral order has little to do with salvation after death. On the contrary, Tarahumaras "ask forgiveness" so that Our Father and Our Mother will continue to bless them in this life. One Rarámuri explained to Merrill that the phrase *wikála tána* does not really mean "to ask forgiveness" but "to ask God for good crops, rainfall, long lives, and protection from sickness" (Merrill 1988:67). Tarahumaras have complex ideas about what happens to their souls after they die, but eternal damnation is not a part of their theology.

The meaning of forgiveness is just one example of a deeper phenomenon. Even though most Rarámuri consider themselves Catholics, their interpretation of Catholic supernatural beings, beliefs, and rituals is radically different from the Catholicism of the missionaries who converted them. Some of the forms may have changed, but the fundamental philosophical assumptions remain grounded in Native American rather than European premises. The complexity of Rarámuri religion demonstrates

Plate 21. Tohono O'odham *wi:gida* figures. The wi:gida ceremony, which includes dancers in turkey-feather headdresses, may reflect the influence of Pueblo peoples on the O'odham at some time in the past, perhaps in late pre-Columbian times. These figures, by Danny Flores in 1991, represent Naviju, left, who taught the O'odham the ceremony, and a clown. Flores learned to carve figures from Hopi kachina carver Cecil Calnimptewa. (Photograph by Helga Teiwes, Arizona State Museum)

Plate 22. Tohono O'odham women picking saguaro fruit in 1854. (Illustration by Arthur Schott)

Plate 23. Juanita Ahil picking saguaro fruit. (Photograph by Walter McQuarry)

Plate 24. Tohono O'odham saguaro wine-making paraphernalia. Top: a saguaro syrup cooking jar, made before 1950, resting on a beargrass and yucca pot ring, made by Rosita Stein before 1963. Right: A wine basket, made before 1916. Lower left: A wooden ladle, made before 1922. Left center: A saguaro-fruit-gathering stick, made before 1962. Left: A sotol strainer basket, made before 1939. (Photograph by Helga Teiwes, Arizona State Museum)

Plate 25. O'odham bowl baskets. They are, top right, a basket made by Nellie Preston, an Akimel O'odham, before 1940; lower right, a Tohono O'odham basket, made before 1941; center, an Akimel O'odham basket with a Man in the Maze design, made before 1950; lower left, a basket made by Margaret Lewis, a Tohono O'odham, before 1970; top left, an Akimel O'odham basket, made before 1929. (Photograph by Helga Teiwes, Arizona State Museum)

Plate 26. Akimel O'odham women in 1854. (Illustration by Arthur Schott)

Plate 27. O'odham pottery. They are, clockwise from upper right, a wedding vase, made by Rupert Angea, a Tohono O'odham, before 1985; a Tohono O'odham black-on-red canteen, made by Susie Miguel before 1959; an Akimel O'odham bowl with encircling figures, made around 1900; and a friendship bowl, made by Rupert Angea before 1985. (Photograph by Helga Teiwes, Arizona State Museum)

Plate 28. A Tarahumara cross and rosary, made before 1978. (Photograph by Helga Teiwes, Arizona State Museum)

Plate 29. A Tarahumara violin, made by José María Nava before 1977. The Rarámuri received European musical instruments—as well as Old World plants, animals, and tools—from the missionaries. (Photograph by Helga Teiwes, Arizona State Museum)

Plate 30. A Tarahumara Fariseo (Pharisee) headdress and sword from Huahuacherare, 1978. (Photograph by Helga Teiwes, Arizona State Museum)

Plate 31. Tarahumara Fariseo dancers on Holy Thursday. The Fariseos are accompanied by drums that are played only during the Easter ceremonies. (Photograph by John P. Schaefer)

Plate 32. A Tarahumara Soldado and Fariseo by the entrance to the church of Huahuacherare, Chihuahua, on Good Friday in 1978. (Photograph by John P. Schaefer)

Plate 33. Tarahumaras dance around a straw Judas lashed to a pole at Huahuacherare, Chihuahua, on Holy Saturday in 1978. (Photograph by John P. Schaefer)

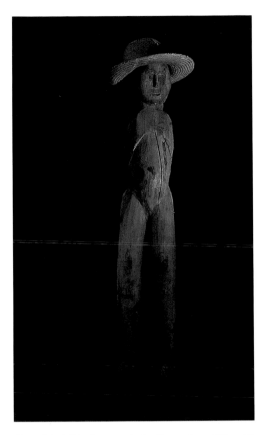

Plate 34. A Tarahumara Judas figure from Rijisuchi, made before 1978. (Photograph by Helga Teiwes, Arizona State Museum)

Plate 35. Southern Paiute basketry water jars, all made before 1919 in southern Utah. Pitch covering the basket kept it watertight. (Photograph by Helga Teiwes, Arizona State Museum)

Plate 37. Southern Paiute wedding baskets. The one at left was made in 1961, and the one at right was made by Edith King of Navajo Mountain before 1970. Southern Paiutes living on or near the Navajo Reservation make these baskets for Navajo ceremonial use, as well as for tourists and collectors. (Photograph by Helga Teiwes, Arizona State Museum)

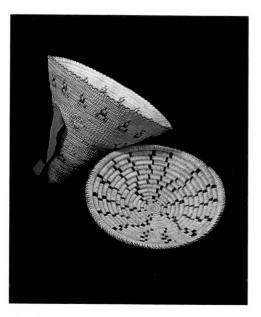

Plate 36. San Juan Paiute baskets. At right is a bowl basket made by Marilu Lehi, and at left is a burden basket made by Maybelle Lehi. Both were made at Hidden Springs in about 1983. (Photograph by Helga Teiwes, Arizona State Museum)

how misleading appearances can be. In the past, anthropologists often tried to determine how "acculturated" particular native groups were by compiling lists of traits and customs, some of which were "aboriginal" while others were of European origin. But the culture of a living people is a dynamic process, a dialogue between outward manifestation and inner meaning. The symbols and patterns of behavior may seem familiar to outsiders, but beneath that familiar appearance a world of difference often exists. Nowhere is the contrast between form and meaning more striking than in the Rarámuri interpretation of Holy Week—the most elaborate, and most "Christian," celebration of the Rarámuri year.

The Tarahumaras call the Holy Week celebration Norírawachi—When We Walk in Circles—because participants march ceaselessly in procession around the church. During the rituals, ceremonial participants are divided into two organizations—the Soldiers (sontárasi) and the Pharisees (pariséo) (see pls. 30–32). The organizations spring from Passion plays, reflecting the influence of Jesuit and Franciscan missionaries. But the reasons the Soldiers and Pharisees dance, beat drums, and march around the church have more to do with Uto-Aztecan than Christian ideas about the order of the universe. According to the Tarahumaras, the universe is arranged in a series of levels, with God and God's wife occupying the highest level, and the Devil (Riré Bitéame) living on the lowest. God and God's wife created the Rarámuri and other American Indians. They are associated with the Sun and the Moon and are represented by Jesus and the Virgin Mary in Catholic iconography. The Devil and his wife, in contrast, created all non-Indians. God is essentially benevolent toward the Tarahumaras, and the Devil inflicts disease and misfortune upon them.

But God and the Devil do not represent absolute good versus absolute evil, as they do in Christian cosmology. The Rarámuri emphasize balance rather than ultimate triumph. Neither God nor the Devil can defeat one another except during Holy Week. During that ominous period, however, the Devil gets God drunk, dangerously weakening him. He also seduces God's wife with his brilliant guitar playing. God and his wife therefore seek refuge in the church, where the Tarahumaras must defend them.

The Soldiers and the Pharisees march because Holy Week is a time of great danger—a time when God is vulnerable and must be protected from the Devil. In William Merrill's words, the Tarahumaras are "God's saviors in the Sierra Madre" (Merrill 1983). If the Devil destroys God, the world will be destroyed as well. The sacred obligation to protect God is stressed again and again throughout the rituals of Holy Week. In the pueblo of Basíhuare, Merrill recorded an oration given by the captain general of the Soldiers on Maundy Thursday:

> Come together here again later today in order to care once more for Our Father who has been drinking recently. For a week he has been drinking but by tomorrow afternoon he will be sober again. So with strength come together to continue caring for him, completing processions around the church. Yes, assemble here again in good spirits later today to continue caring for Him. Do not be lazy in your care; later on everyone can rest. In this fashion, we will be asking for many more years to live. Do not become disheartened, I say to you. Everything I am saying to you is the word of Our Father. (Merrill n.d.)

Syncretism and Survival

Ritual expresses the moral order of a society, defining and reinforcing its deepest values and beliefs. It is also one way a people define themselves in relation to others. The Rarámuri have survived for the past four hundred years by pursuing a variety of strategies, including military resistance and geographic retreat. But they have also adapted to changing political and economic forces by incorporating foreign elements into their culture and making them their own. Sheep, goats, Catholic ceremonies—all have enriched and strengthened Tarahumara society. "In a very real sense," anthropologist Robert Zingg noted, "the Tarahumaras live on the sparse grass of the mountainside, collected for him by the animals and transformed through the cycle of animal food to a fertilizer for the corn which is the staff of Tarahumara life" (Bennett and Zingg 1935:13). Old World animals nourish New World plants. Such a syncretic economy—one that blends

THE RAILROAD AND THE TOURISTS

Thomas E. Sheridan

In 1961 the Ferrocarril Chihuahua al Pacífico did for the Rarámuri what the Atchison, Topeka & Santa Fe had done for the Hopis and Navajos nearly a century before. The railroad, which runs from Chihuahua City to Topolobampo near Los Mochis on the Sinaloa coast, sliced across the Sierra Madre Occidental and breached the isolation of the Tarahumaras with iron rails. One result was the expansion of the tourist industry in Tarahumara country. Another was the commercialization of Rarámuri arts and crafts.

Like Navajo blankets and Hopi pots, these arts and crafts originated as utilitarian or ritual items. Women produced coil-and-scrape pottery and twilled baskets of beargrass, sotol, and pine needles. During pre-Columbian times they also wove blankets of agave fiber or Indian hemp (*Apocynum* sp.), and woolen textiles after the Spaniards introduced sheep and goats. Men fashioned musical instruments such as double-headed snare drums, cane flutes, and cocoon leg rattles of pre-Columbian origin and carved wooden violins patterned after European models.

Beginning in the 1960s, however, Mexican merchants and Jesuit missionaries like Father Luis Verplanken encouraged the Rarámuri to manufacture textiles, pottery, baskets, and wooden animal and human figurines to sell in local shops along the railroad. Traders also descended on Tarahumara households to purchase those items for resale in cities across the United States and Mexico. Gallery shows and museum exhibits, including a major exhibit at the Arizona State Museum in 1979, further stimulated the demand for Rarámuri goods.

By the 1990s, Tarahumara craft production had developed into a cottage industry in the Sierra Madre. Meanwhile, more and more tourists were visiting the Rarámuri themselves, staying at hotels along the railroad or rolling into Tarahumara communities by the busload to view the ceremonies of Holy Week. As with other such developments, the effects have been mixed. Tarahumara governors have attempted to regulate the behavior of the tourists to prevent them from disrupting the rituals, but the control of Rarámuri craft production has been harder to achieve. Jesuits like Verplanken and trader-anthropologists like Barney Burns have worked hard to preserve the quality of Tarahumara items and to ensure that the Tarahumaras receive a fair price for their products. Other traders and merchants lack such ethics and pay a pittance for Tarahumara carvings, pots, and baskets. A few experiments at mass production have been attempted, and Mexican craftspeople in the Sierra Madre are beginning to churn out "Tarahumara" items in the same way that hundreds of Mexicans along the Sonoran coast are imitating Seri ironwood carvings.

In part because of the railroad, the Rarámuri are under assault—from Mexican loggers, ranchers, farmers, and drug runners, and from unscrupulous traders who seek to market "exotic" Tarahumara culture while raking off the profits from the commercialization themselves. Neither tourism nor craft production is intrinsically good or bad. Their ultimate impact will depend on the ability of the Rarámuri to control these developments on their own terms.

both Europe and the Americas—allows the Rarámuri to flourish in their mountain strongholds.

The celebrations of Holy Week do the same. The Norírawachi ceremony must have reassured and delighted the missionaries of the past, just as they do Mexican observers today. Most Tarahumaras, after all, are bautizados (*pagótame*)—Christians, albeit colorful ones, like everybody else. But the message conveyed by the Tarahumaras themselves is much more subtle and complex. We are Rarámuri. We are distinct. We must restore balance. During Norírawachi, God himself needs our help.

That resilience has been tested again and again. During the colonial period, the Spaniards seized the fertile valleys of the eastern foothills and forced thousands of Tarahumaras to work in their mines and haciendas. During the nineteenth century, the Reform Laws of the 1850s prohibited the communal ownership of land, and non-Indians wrested control of good farmland away from the Rarámuri, provoking violent revolts in Nonoava (1876), Guadalupe y Calvo (1895), and Chinatú (1898). And during the reign of Porfirio Díaz, who ruled Mexico from 1876 until the Mexican Revolution in 1910, foreign mining and lumber companies penetrated many parts of the Sierra. They even built a railroad that ran from Chihuahua City to the new Mexican community of Creel, right in the heart of Tarahumara country. According to Merrill, "Only the disruptions accompanying the Mexican Revolution prevented the wholesale destruction of the forests" (1988:39).

The twentieth century brought paternalism as well as greed. The Jesuits returned in 1900 to build churches, clinics, and schools. After the Mexican Revolution, the federal government also launched a number of schools and "cultural missions" to integrate the Rarámuri into Mexican society. In 1952 the National Indian Institute (Instituto Nacional Indigenista, or INI) established a coordinating center at Guachochi. One of its major goals was to protect Tarahumara lands by organizing the Tarahumaras—and their *mestizo* neighbors—into *ejidos*, the federally chartered peasant groups that hold land and water in common. All of these programs attempted to better the material conditions of Tarahumara life and educate Tarahumara children. And while many tried to preserve certain aspects of Rarámuri culture, particularly language and arts and crafts, the basic thrust was assimilationist. No agency, public or private, was content to leave the Rarámuri alone.

That would have been impossible anyway. Thousands of *chabochis* (non-Indians, "whiskered ones") were moving into Tarahumara territory, growing crops, running cattle, opening stores, and cutting down trees. Merrill notes that in 1900 the number of chabochis in the Sierra was only slightly higher than the number of Indians but that the population of non-Indians was three times higher in 1940 and six times higher in 1980 (Merrill 1988:41). Non-Indians now dominate many of the ejidos to which the Tarahumaras belong. During the twentieth century, the Rarámuri have become a minority in their mountain homeland.

The pressures have grown even more intense, and more dangerous, in recent years. With the explosion of the drug trade, the hidden canyons of the Sierra have attracted many *narcotraficantes* (drug runners). Indian farmers are often caught in the middle, becoming pawns in the deadly game between the narcos and the Mexican army. Logging has also increased, bringing more roads and more erosion as stands of timber are clear-cut. With Mexico's population soaring, more and more hands are trying to pry loose the resources of the Sierra and tear them away from the Rarámuri.

The Tarahumaras have often paid a terrible price for their survival—silver mining and repartimiento during the colonial period, drugs and logging today. Outsiders have long predicted their demise, just as they have been predicting the demise of Indians across North America for the last five hundred years. But the Rarámuri continue to cling to their homeland with tenacity and skill. They have fought with weapons, and they have fought by receding into the mountains and canyons. They have taken what they wanted from the outside world and used it to strengthen their own way of life. Rarámuri society is flexible; Rarámuri culture is creative and resilient. People who see themselves as God's protectors will not easily disappear.

THE NɨNGWɨ (SOUTHERN PAIUTES) THE PEOPLE OF THE NORTHWESTERN FRONTIER

Catherine S. Fowler and Nancy J. Parezo

The wooded plateaus overlooking the Grand Canyon on the Arizona-Utah border constitute a cultural frontier between the Great Basin and the desert Southwest. Here live the San Juan and Kaibab Southern Paiutes. In the low deserts flanking the banks of the Colorado near Needles, California, live another Southern Paiute group, the Chemehuevis. Since the 1770s, when Spanish explorers first encountered these societies, Anglo and Hispanic scholars and settlers have remarked on their cultural stability and tremendous environmental knowledge. Yet the Southern Paiute homelands have always been areas for contact and exchange with the peoples who inhabited the high arid deserts of Nevada and Utah, the Upland and River Yumans, the Utes of southern Colorado, the Navajos, and the mesa-dwelling Hopis, whom Southern Paiutes believe are their ancestral neighbors.

The Southern Paiutes call themselves Nɨngwɨ, which means person or speaker of my language. But they also recognize small social groupings based on the areas they inhabit—the Kaibab, Las Vegas, Moapa, San Juan, Shivwits, and Uinkarets, for example. The Kaibabs call themselves *kaivavitunɨngwɨ* (Mountain Lying Down People; Kaibab Plateau); the Las Vegas, *nivaganicinɨngwɨ* (Snowy Mountain People; Mt. Charleston); the Moapa, *mudinɨngwɨ,* (Bean People; along the Moapa River in southeastern Nevada); the San Juan, *tuyouipinɨngwɨ* (Rock River People; northern Arizona south of the San Juan River); the Shivwits, *sipicimɨ* (People Who Live to the East; in northwestern Arizona north of the Colorado); and the Uinkarets, *yɨpinkat-*

itɨcimɨ (People of Ponderosa Pine Peak; northwestern Arizona north of the Colorado). The Chemehuevis call themselves *tantawats* (Southern Men), *tuumontcokowɨ* (Black Bearded Ones) or merely *nɨwɨwi* (the People) but distinguish between those who live in the mountains from those who live along the river.

The Children of Coyote

All Southern Paiute peoples see themselves as related to the children of Coyote. Human beings came about through the union of Coyote and the daughter of Ocean Woman (Water Woman), who lived far to the west. She is credited with the creation of the earth, after which Coyote made animals of mud and people. According to others, Wolf made the land and animals. All agree, however, that Coyote then carried the Nɨngwɨ to their homelands.

According to Chemehuevi George Laird, after Coyote and the daughter mated, Ocean Woman placed the new human beings in a basket—a narrow-necked water jar—and told Coyote to carry them far away to a distant land, a good place where they could hunt and gather the foods they needed. Ocean Woman told Coyote not to open the jar under any circumstances until he came to a good place. Coyote agreed because he had been gone from home for a long time and wanted to visit his brother, Wolf.

Coyote began his journey, but the basket was heavy. As he walked along with the water jar on his back, he heard sounds coming from the jar. He was very curious—as coyotes always are—and wanted to see inside. Finally Coyote could stand it no longer and opened the jar. People came tumbling out in every direction! First were the coast people, and Coyote named them as they came forth. Then came the adjacent tribes. These were the Pasagʷagaditsɨ (Blue-Green-Water Dwellers; the Havasupais), the Waddipayatsɨ (Hualapais), the Ayate (Mohaves), the Kwitsanɨ (Quechan), the Hatpah (Pimas), and the other peoples who live nearby, including the Hukʷantsi (Europeans). These people began traveling in different directions.

Coyote quickly closed the jar, but only a few people were left. These were the Nɨngwɨ—the Southern Paiutes. He took them to his older and wiser brother, Wolf, who scolded Coyote for disobeying. Opening the jar, Wolf and Coyote molded and named the Nɨngwɨ as special people. The brothers carried them to where the *nanta*, or agave, grows, to where the deer and the mountain sheep are plentiful—to the edge of the great canyon overlooking Paga, the awe-inspiring Colorado River. There they remain to this day, following the ways Coyote taught them.

The Southern Paiutes: A Stable Desert Adaptation

When Coyote first brought the Southern Paiutes to the area north of the Colorado River, they were hunters and gatherers who efficiently used all the available foods and resources. They hunted large and small game and various birds, using bows made of juniper and bent bighorn sheep horn, and arrows of cane, service berry, or arrow weed. Some groups obtained fish from the few rivers and lakes in the plateaus and desert basin. During lean years, the Nɨngwɨ also ate a variety of nutritious insects, which could be taken in large quantities when they swarmed. These food habits were later misunderstood and disdained by Anglo American travelers because they were so foreign. This was only a difference in perspective, however. According to one Chemehuevi woman, "To the white man, the desert is a wasteland; to us it is a supermarket" (quoted in Laird 1976:5).

The Southern Paiutes collected and processed more than a hundred species of plants that varied by altitude: acorns, pinyon nuts, cactus fruits and stems, yucca fruits, mescal crowns, biscuit roots, sunflower seeds, berries, and seed grasses. This complex diet required a tremendous knowledge of place and exact timing, for each plant was available for only short periods each year.

Gathering and processing plant foods was largely the responsibility of women. Seeds were swept with a basketry beater into a conical container, winnowed, parched, and then ground to a flour with a stone mano and metate. Women heated pebbles and dropped them into a cooking basket along with the flour to produce mush. The flour also was baked as bread in the ashes of campfires. Food was eaten with horn spoons and ladles.

The Southern Paiutes' hunting-and-gathering lifestyle and the varied but harsh terrain meant that they had to use materials from both the lower desert and the high plateau, which is stud-

Mabel Lehi, a San Juan Paiute, using a seed beater in 1984. (Photograph by Susan Brown McGreevy)

ded with mountains and deep canyons. To do this, they moved in a seminomadic pattern, often traveling considerable distances to take advantage of resources that changed each year. Each band always had alternative food sources, knowledge about which was passed on from generation to generation. Fall was the time of plenty, when families made trips to the plateaus to collect yucca fruit, harvest pine nuts, particularly from single-leaf piñon, and hunt deer. Stores were cached for the winter but often ran short; late winter and spring could be times of famine.

Each summer Nɨngwɨ families returned to their home base at the foot of the plateaus. The availability of drinking water determined the base's location, as well as that of their temporary living sites, because streams and springs were intermittent. Families had use rights to springs, but this did not mean that others could not make use of them or the surrounding area on request. Southern Paiute behavior was, and still is, predicated on an ethic of communalism and sharing. As Captain George told anthropologist Isabel Kelly, "If someone else came around, he could camp there too; a man liked to have company. He liked to move around and change springs too. He knew where to camp" (1964:7). It was the headman's responsibility to decide when to move camp to the next spring. So important were these springs that each had a recognized name, known to all group members.

Some areas of Paiute country functioned as communal lands. The flats near the springs on the Kaibab Plateau, for example, were available to anyone who wanted to harvest seeds. People would gather in these areas during the fall and winter to conduct a rabbit drive and to hunt deer. According to Kwagunt, a Kaibab Paiute who lived with his family in House Rock Valley, rabbit hunts were highlights of the year. In 1938 he gave the following description of a rabbit hunt to B. A. Riggs:

> We would find a place where rabbits like to run and set up our nets. The nets were made of yuka [yucca] strings. They were from ten to twenty-five yards long. We would put all we had out in a long string, sometime a hundred yards long. We would put all them up on stakes driven into the ground. When we got our nets set we would make a circle out around the part of the country we wanted to make our drive through. We then formed into a half circle a short distance apart and would go towards the net. The rabbits would go to the nets and follow along it. There would be an Indian stationed at each end and sometimes along in the center. When the rabbits came along the Indians would shoot them. We would take the rabbits to camp, build a large fire out of sagebrush, and would take enough rabbits to make a meal for the camp and put them in the fire and burn the hair off, when we didn't want the hair to make robes with. When the hair was all burned off we moved the coals and ashes away and put the rabbits in a pile, then bury them in hot ashes and coals. When they are cooked we take them out and pull the ears off and give them to the older people, chief or Medicine man. That was the best part of the rabbit. The children got a leg or piece of the back. (Euler 1966:87)

In winter, women wove rabbit skin into robes and made twined-bark leggings for warmth. At other times, Southern

Paiutes wore "pieces of buckskin hanging from their waists," according to Franciscan missionary Francisco Atanasio Domínguez in 1776. Travelers in the 1870s noted more elaborate clothing—fringed bark or skin shirts, basketry caps and sandals, and breechcloths and hide shirts—as well as the continuing use of rabbit skin robes. In 1933 Isabel Kelly made a collection of twined-bark clothing from the Kaibab and Shivwits Paiutes, including yucca sandals, which, she noted, "quickly gave out if you walked around much."

Nɨngwɨ housing varied by season, and—like their woven mats and bags, basketry cradles, and pipes—was made of local materials and changed little over the centuries. The winter house was constructed of bent willow boughs planted in the ground, thatched with cedar bark, and often covered with earth. These semiconical houses, similar to Western Apache wickiups, were temporary structures designed for only one season. They were owned by both the man and the woman since they were the product of their joint labor. In summer and fall only shades and circular windbreaks were used. After 1900, canvas-covered tipis, similar to those made by the Utes, became common.

The Southern Paiutes' economy and technology were remarkably stable for many centuries, but this does not mean that their life styles did not change. The Nɨngwɨ have often looked toward the north for new ideas. The Utes are particularly respected, because, as one individual recalled, "they know the Bear and the Sun dances. They know how to make buckskin dresses and gloves and how to make beadwork" (quoted in Kelly 1964: 33). Kaibab Paiutes acquired tipis, travois, buckskin clothing, beadworking techniques, skin-covered cradleboards, a braided hair style, and the Bear Dance from the Utes. The Nɨngwɨ also had extensive trade contacts with their neighbors to the south and east, and their ideas and materials have been enriched as a result.

Agriculture is an excellent example of this process. By 1776 some Southern Paiute groups were farming, a practice that they probably learned from the Hopis or Mohaves. One Paiute word for squash seems to have been borrowed from Hopi (*parangwar* in Paiute and *patnga* in Hopi). Agriculture was easily integrated into the seasonal food-gathering cycle. Southern Paiutes used a

The summer camp of Kaibab Paiute Chu-ar-um-peak and his family in 1872. They were met by John Wesley Powell as he explored the Grand Canyon region. Powell told his readers, "They have never received anything from the government and are too poor to tempt the trader, and their country is so nearly inaccessible that the white man never visits them" (1895:318). (Photograph by J. K. Hillers, Arizona Historical Society)

variety of techniques to increase the quantity and availability of wild plants. They burned brush and grassland to stimulate growth and deliberately propagated mesquite and wild grapes by transporting seeds to suitable locations (Stoffle and Dobyns 1983). In 1776, while in the valley of the Virgin River, the Span-

SOUTHERN PAIUTE BASKETS

Nancy J. Parezo

Until the early twentieth century, basketry was much more important than pottery in Ningwɨ technology. Fibers of different plants—willow, acacia, cottonwood, sumac, and devil's claw—were used to make twined and coiled baskets, each designed to fill a special need (see pls. 35–37). The Southern Paiutes employed flat, round baskets for parching seeds and for gambling games; fan-shaped trays for winnowing; large, conical burden baskets with leather straps for carrying gathered and domesticated foods; coiled food bowls for eating; globular constricted-neck baskets for holding keepsakes; a large, circular coiled tray for parching beans with hot coals; and a spoon-shaped seed beater to knock the ripe seeds from standing grass stems. The Ningwɨ needed baskets to process and store all kinds of plant foods and even used basketry techniques to make sandals, carrying straps, hats, cradleboards, narrow-necked water jars, and caps. Women spent much of their time producing and repairing baskets.

The Southern Paiutes and their neighbors have exchanged ideas about basketry and may have traded baskets for a long time. In the past, baskets produced by Ningwɨ groups in northern Arizona and southern Utah were very similar to Havasupai and Hualapai baskets, while those made by Chemehuevis were similar in size, shape, design, and technique to Mohave baskets. The Havasupais, Hualapais, and Southern Paiutes, for example, finish the edges of their burden baskets with an under selvedge of at least one row of three-strand twining and then a rim coil over a bundle of bent warp rods, a technique rarely seen among other peoples. A method of starting twined baskets by crossing two pairs of rods is also shared, as is the general technique of diagonal twining. Even with these similarities, Upland Yuman and Southern Paiute twined baskets can be distinguished because the Ningwɨ kept the outer, rounded surface of the split fiber element on the outside of the basket, while Havasupai and Hualapai weavers kept it on the inside.

Chemehuevi coiled baskets are noted for their restrained, simple designs; fine, even stitching; and graceful contours. Baskets produced before 1900 had simple geometric designs in black with large undecorated areas. Designs were placed in panels, spirals, or banded blocks but were never extended to the bottom of the basket. Occasionally, small trade beads were woven into the basket to enhance the design. Contemporary work is more complex, featuring life forms such as birds, butterflies, coyote tracks, snakes, lizards, bugs, and plants. Anthropomorphic motifs, however, have never been popular. Designs are thought to be the personal property of the weaver, although a weaver might share them with relatives. Mary Snyder is credited with creating the diamond-backed rattlesnake and the bug basket designs (two of the most recognizable of all Chemehuevi designs) in the late 1890s. Other weavers—such as Maria Painter, Nellie Fisher, Mary Smith, and Mary Chappo—also made distinctive designs in individual styles that served as trademarks of their work.

Historically, most Southern Paiute coiled baskets were undecorated, but in the 1870s the baskets that John Wesley Powell collected from San Juan and Kaibab Paiutes for the Smithsonian Institution were decorated with a dark band and

A Kaibab Paiute basketmaker in 1873. (Photograph by J. K. Hillers, Smithsonian Institution, National Museum of Natural History)

made with a three-rod foundation. Both features are typical of Havasupai and Hualapai baskets. Today some basketmakers are reviving the old style in order to preserve individual and collective memories of the old ways. Individuals like Marie Lehi and her daughters and granddaughters are exploring new designs and colors. San Juan Southern Paiute basketry is currently in a period of great expansion.

Contemporary Chemehuevi baskets are made of willow, sumac, and occasionally cottonwood, with designs executed in black devil's claw, tan and gold juncus grass, red yucca root, or red-pink split flicker quills. Like other modern Paiute basketmakers, Chemehuevi weavers use brilliant colors, dying sumac and willow splints in commercial solutions. They also use their traditional knowledge of plants in order to produce vegetal dyes. It is not uncommon for contemporary weavers to include three or four colors in a basket. They use a three-rod triangular foundation worked from the right, a distinguishing feature of Chemehuevi baskets. Like other groups, Chemehuevi weavers use devil's claw on the braided rim to provide strength as well as color contrast.

ish explorer and missionary Silvestre Vélez de Escalante noted in his diary that on the small plain and on the bank of the river there were three small corn patches with very well made irrigation ditches. He also noted that on the mesas on either side for a long distance lived Indians who sustained themselves by planting maize (corn) and calabashes (squash). In the early to mid 1800s, the Nɨngwɨ were growing corn of several colors, white and brown tepary beans, squashes, and white-seeded amaranth, a favorite plant of Yuman and northern Mexican peoples. The Southern Paiutes also grew Spanish-introduced wheat, while the Chemehuevis grew potatoes, muskmelons, and cotton. Most bands irrigated little, depending instead on summer rainfall.

The harsh environment precluded large groups from remaining together on a year-round basis, thus eliminating the necessity of creating complex social and political forms. While the Southern Paiutes have always thought of themselves as Nɨngwɨ, they have not, until recently, worked as a tribe. Instead, the basic social groups were individual extended families and groups of related families who occupied a given territory and who were referred to by a significant geographical feature in the area. Families developed an intimate knowledge of their well-defined homelands and devised many ingenious techniques to utilize its sparse resources.

Nɨngwɨ family groups occasionally came together on social occasions. One of the most important was the Round Dance (*nɨkkappɨ*), a harvest dance in which men and women, dressed in their finest clothes, beads, and face paint alternated in a circle facing inward with hands joined. Dancers moved in a clockwise direction while men took turns singing a capella or to the accompaniment of a rasp and rattle. Today, large groups of Southern Paiutes and their neighbors gather for "big times," to sing and dance. The Bear Dance, introduced from the Utes in the early 1900s by Tom Moustache and Mampuc, is one of the most important of these events.

Rather than a single head chief, the Paiutes recognized local leaders who were selected with a kin group's popular consent. These headmen were individuals noted for their hunting expertise, knowledge of the land, and speaking ability. For example, one Paiute man noted that "Keno spoke early in the morning, every morning, standing by the door of his house. He spoke loudly so all could hear from their camps. He told the people how to hunt and where to hunt; and after a time everybody answered, 'Yes.' Then they went for deer" (quoted in Kelly 1964:27). At other times a headman would advise the people. Captain George said that Tapič, the leader of one Kaibab group in the 1850s and 1860s, was a wise man because "he told the people to be good; to be kind to each other and to share their food: when they had deer or rabbits to give half to other camps" (Kelly 1964:28). The headman's wife led the women in gathering seeds and was responsible for announcing the name of the new headman when the incumbent was considered to be too old. The headman served as an advisor and had no power to enforce his views on others or to control other groups. If the people respected his wisdom, he was followed.

Nɨngwɨ life was, and is, aesthetically rich in songs and stories. A favorite time for telling legends was in the evening during the winter months. Narrators risked snakebites if they related stories during the summer. Elderly men would recite far into the night. Most oral literature is set in the Time When Animals Were People, a distant past when animals had the power of speech and established customs for people through their adventures. Also common are the stories of the origin and nature of the world, in which the elders explain that the stars are "the Indians' watch" (Kelly 1964:137), how time and the seasons work, that the rainbow is the rain's walking stick, and that every person has a soul, which travels west at death.

Religion and ceremonies centered around the spirits of animals and plants, springs and other water sources, mountains and valleys, the heavens, and the earth in general. The Southern Paiutes, like their neighbors the Northern Paiutes, have emphasized curing in their religion in the nineteenth and twentieth centuries. A doctor or shaman, called a "dreamer," was "usually an old man, sometimes a young one, but never a young woman," according to Captain George and Sarah Frank, Kaibab Paiutes who worked with Isabel Kelly in the 1930s (1939:151). Dreamers were skilled healers who could also predict where food or lost objects might be found. They were respected individuals associated with the locality in which they lived. For example,

Yinimu of Tinka'nivats, and Tcanga and Mi'mitana of Short
Creek were noted healers at the turn of the century. Some shamans had special powers. Tuku'nimp of Navajo Wells was a
famous rattlesnake healer in the late nineteenth century. His
specialty allowed him to cure snakebites by virtue of powers
derived from the spirit of rattlesnakes.

These powers came unsought to healers in dreams from a
supernatural tutelary—an animal (such as bear, coyote, wildcat,
or eagle) or a human—who taught the dreamer how to cure
through song and medicine. According to Little Jim of the Shivwits Paiutes, disease is caused by intrusion, either of disease or
from the actions of a ghost, and it is the responsibility of the
dreamer to remove the intrusion through proper curing techniques and ritual (Kelly 1939:155). A wide range of illnesses,
symptoms, and curing techniques were known in the late nineteenth and early twentieth centuries.

History: Avoidance and Contact

Like other groups in the Southwest, Southern Paiutes identify
strongly with their land. Their beliefs are rooted in and sanctioned by their origin stories. Some anthropologists believe—
based on studies of language, myths, and prehistoric artifacts—
that the Nƚngwƚ originally came from southern California and
the Pacific coast. They spread across the Great Basin and into the
Southwest sometime after A.D. 1000. The camps of Southern
Paiute ancestors can be recognized by distinctive pottery, manos
and metates, and brush wickiups that differ from those of con-
temporaneous Western Pueblos. Southern Paiute pottery has
been found at sites in the Moapa Valley of Nevada dating from
around A.D. 1200. Evidence of Kaibab and San Juan Paiute ancestors in their homeland dates to A.D. 1300.

Although the Nƚngwƚ were probably aware of the presence of
the Spaniards long before they actually saw them, we know little
of this interaction until roughly 1776. In March of that year, an
expedition led by the Franciscan missionary Francisco Garcés
marched up the Colorado River and made contact with the Chemehuevis. At that time the Chemehuevis lived in the deserts
west of the Mohaves rather than along the Colorado River.

MARKING THE TRADITIONAL LANDSCAPE

Nancy J. Parezo

San Juan Paiutes feel they stand in a special relationship to that part of the natural world that they see as their homeland. This relationship is considered sacred and is specified in their religious beliefs, oral history, and world view (Stoffle and Dobyns 1983). San Juan Paiutes teach their children about this relationship, which includes treating the plants and animals found in these areas with special reverence as beings with spirits, and gathering medicinal plants with care and a respectful attitude.

Each Ningwɨ community has a knowledge of sacred places within its local territory which are highlighted in local versions of the traditional origin myths. For the San Juan Paiutes, for example, Coyote untied the sack (jar) from which people escaped near Page, Arizona. This place they consider the original home of all Paiutes. Also included are the exact location, history, and character of the kinds of sites that the San Juan Paiutes regard as sacred: the graves of ancestors, all sources of water (springs and wells), and Round Dance sites. Even localities that have not been used recently, due to Navajo occupation, are distinctly remembered in oral-history accounts and stories (Bunte and Franklin 1987).

Garcés's diary provides an account of this meeting, describing the crooked staffs carried by Chemehuevi men as well as their recurved bows and caps decorated with quail crests. These features were also illustrated nearly a century later by Heinrich B. Möllhausen, the artist who accompanied Lieutenant Joseph C. Ives on his topographic reconnaissance of the Colorado River in 1857 and 1858.

In July 1776 a second Spanish expedition, that of Fray Atanasio Domínguez and Fray Silvestre Vélez de Escalante, left Santa Fe and made a giant arc through western Colorado, central and southern Utah, and northern Arizona before returning to Santa Fe by way of the Hopi and Zuni lands. Along the Virgin River north of the Grand Canyon, they contacted several groups of seed-gathering peoples whom they called Payuchis. Although the Ningwɨ shunned contact, Escalante described them as individuals who ate grass seeds, hares, pinyon nuts, and dates and in some places also cultivated the soil, growing corn, beans, and squash.

The Spanish presence was felt more directly and intensely later in the century as a result of the slave trade. Spanish settlements in New Mexico, northern Mexico, and southern California sought forced labor for various enterprises, and the traffic in slaves was extensive. The Southern Paiutes were in the unenviable position of living between Ute slave raiders on the north and Navajo raiders on the south, both of whom preyed on Nɨngwɨ children and young adults. By 1810 many Spanish raiders were working along the Old Spanish Trail, a major trade route that linked Santa Fe and settlements in California. In 1839 trader Thomas Farnham reported that Southern Paiutes were "hunted in the spring of the year, when weak and helpless, by a certain class of men, and when taken, are fattened, carried to Santa Fe and sold as slaves during their minority. (A likely girl) in her teens brings often-times sixty or eighty pounds. The males are valued less" (Farnham 1843:11).

Slave raiding significantly reduced Southern Paiute populations and sent families into hiding, keeping them away from many of the richest areas within their territory, including those most suitable for farming. The early 1800s were thus a time of hardship, of which little has been recorded in historical documents. It was not until 1859 that descriptions of Southern Paiute groups were penned following the first intensive exploration of the dry and forbidding Nɨngwɨ territory. U.S. Army officers exploring in the vicinity of Marsh Pass and Tsegi Canyon noted that a Southern Paiute family lived near good grazing land and that others lived to the west of Glen Canyon and on the Shonto Plateau. Writers described the San Juan bands as a hunting-and-gathering people organized in small migratory bands of related kinspeople who were peacefully disposed toward or "timid" around strangers. In regions less decimated by slavery and disease, Nɨngwɨ bands were less retiring. They occasionally attacked explorers, traders, and emigrants to California in order to protect their territory.

Settlers eventually came in the 1850s, and Southern Paiute bands struggled to maintain their lifestyle as Mormons filtered into the lands along the Arizona-Utah border. Because the Nɨngwɨ lacked centralized authorities like chiefs or even a concept of themselves as a unified tribe that "owned" territory,

Mormon farmers were easily able to appropriate Southern Paiute lands. Mormons felt superior to the Nɨngwɨ and did not recognize or acknowledge that hunters and gatherers could possess the land. The Mormons therefore commandeered the productive pockets of land, effectively eliminating most Paiute farming. Before long, traditional food supplies were depleted by livestock, timbering, and other Mormon activities.

The Southern Paiutes tried to remain apart from the Mormon settlers, but it was difficult. Because of sustained contact with settlements by the 1870s, some families formed larger and more permanent residence groups in proximity to emerging Mormon towns and began to supplement their subsistence strategies with transitory wage work, part-time exchange labor, the sale of buckskin gloves and baskets, begging, and menial service jobs (Knack 1980:12). Many Nɨngwɨ incorporated new items into their material culture as a result—brass buckets, metal-tipped arrows, and cast-off Anglo clothing—yet they retained their traditional ideas about life. Few actually converted to Mormonism even though Mormons considered it their moral duty to "save" local Indians and teach them "civilized" ways because they considered American Indians to be Lamanites, one of the lost tribes of Israel. Sporadic conflicts with settlers occurred throughout the area, and there was a short-term increase in intertribal warfare between Southern Paiute groups and the Mohaves, Hualapais, and Navajos, all of whom now owned firearms. In some areas, such as on the Arizona Strip, major intertribal conflicts were averted by the activities of Mormon missionaries like Jacob Hamblin, "the Buckskin Apostle," who was active throughout much of southwestern Utah and northern Arizona from the 1850s through the 1870s.

Southern Paiutes soon encountered difficulties with the federal government. Their isolated region became part of an administrative "frontier" located between agencies in Arizona and Utah, neither of which took responsibility for Nɨngwɨ welfare and lands. Tribal status fluctuated with changes in government policies. In the 1870s the government made some attempts to settle the Nɨngwɨ on reservations so that the people would not be a "serious burden" to settlers. In 1872 the Moapa Reservation was established with the idea that all the Nɨngwɨ in Utah, Ari-

The baptism of a Southern Paiute man by a Mormon missionary in southern
Utah in 1875. (Utah State Historical Society)

zona, and Nevada would voluntarily go there. However, most
families refused to abandon their homelands. In 1873, John
Wesley Powell and G. W. Ingalls traveled to Utah for the Bureau
of Indian Affairs (BIA) to investigate the needs of the Southern
Paiutes prior to their forced removal to Moapa. After speaking
with a respected Kaibab headman, Chu-ar-um-peak (or Frank),

Powell noted that the people were peaceful and industrious but impoverished—"half-starved and scantily clothed, they present a pitiful, abject appearance" (quoted in Euler 1966:83). Powell and Ingalls recommended assimilation and the establishment of several reservations, for they reported that, in their opinion, the Paiutes "fully understand that the settlement of the country by white men is inevitable. . . . [T]heir hunting grounds have been spoiled, their favorite valleys are occupied by white men, and they are compelled to scatter in small bands in order to obtain subsistence " (Powell and Ingalls 1874:41–42). In the late 1880s, the Shivwits Reservation was established west of St. George, Utah, for the Shivwits and Southern Paiute peoples from the Arizona Strip. Through this act, Southern Paiutes lost most of their territory north of the Grand Canyon. They have never been able to regain it.

By the first decade of the twentieth century, many changes had occurred in Southern Paiute life. No longer able to gather and hunt freely and with their food resources greatly depleted, the Ningwɨ led a precarious life. The Kaibab group from Moccasin Springs was forced into a nomadic existence, unable to keep their annual rounds. Because of Anglo concepts of personal property and trespass, access to water and gathering places was denied. As explorer and artist Frederick Dellenbaugh noted in 1903, the Paiutes could not obtain buckskin, and their whole method of living had changed. For Dellenbaugh, all that remained of the old ways he had seen in the 1860s were some rickety wickiups. Men and women were willing workers when they could find employment, but many had to scavenge around white settlements. Kwaganti, a Kaibab Paiute who knew Powell and Dellenbaugh, moved to Kanab in the early 1900s and chopped wood for settlers. Every year he returned to his spring in Kwagunt Valley to harvest sage seeds and mescal (Kelly 1964:19). Not everyone followed this pattern; some tried to follow the old ways. In the 1930s, Saitimpi and his brother Kwiuinimpi, two widowers, wintered just below the rim of the Grand Canyon and in the spring and summer lived on the plateau near the mouth of Kanab Canyon, where they gathered seeds (Kelly 1964:20). This lifestyle was not possible for most, however. By 1910, when anthropologist Edward Sapir studied their language and customs, few Southern Paiutes were making traditional crafts, and many activities were no longer practiced.

The federal government exhibited no consistency in its treatment of the Southern Paiutes, and each band fared differently. By 1900 about a hundred Kaibab Paiutes lived southwest of Kanab, Utah, and in 1907 roughly 20,000 acres were set aside for them near Moccasin, Arizona, by executive order. Mormon church officials negotiated a deal with Anglo squatters at Moccasin Springs, stipulating that one-third of the water be delegated to the Kaibabs, who could then reestablish a small farm (Euler 1972:84). The BIA issued cattle, and the Paiutes had established a successful herd by 1930. But ranching required more land and water than was available. In 1913 and 1917 the land base was expanded to 120,413 acres. Unfortunately, the reserves were in high desert areas that were inadequate to support the population assigned to them. As a result, many families continued the marginal, town-dwelling pattern that had emerged in the late 1800s. The Southern Paiutes living in towns were not recognized by the BIA.

At the same time, the small band of San Juan Paiutes who lived south of the Colorado River were denied tribal status. Nɨngwɨ in the Hidden Springs and Navajo Mountain areas had lived in contact with Navajos since at least 1864, when Navajo families took refuge with them to escape imprisonment at Bosque Redondo. Southern Paiute lands, engulfed by the expanding Navajo population, were incorporated into the Navajo Reservation in 1884 without being given separate status. In 1907, Nɨngwɨ lands north of the Arizona border and south of the San Juan River in Utah that had briefly been removed from federal protection were affirmed as the Paiute Strip Reservation. Jurisdiction for the region remained with the Western Navajo Agency in Tuba City. The San Juan people again lost their Paiute Strip Reservation in 1922, only to see it later be reincorporated into the Navajo Reservation. Southern Paiutes continue today to occupy areas north and west of Navajo Mountain and to hold traditional sites north and west of Tuba City.

In 1954, four of the relatively small Southern Paiute bands in Utah were terminated from federal trustee status by federal law (68 U.S. 1099). The Nɨngwɨ called these people the Five Bands

Kaibab Paiutes meet with Major John Wesley Powell in 1873. An explorer, scientist, and organization builder, Powell was the first anthropologist to study Southern Paiute lifeways. At their meeting on September 19, Powell told Shivwit leaders, "I wish to learn about [your] canyons and mountains, and about [yourselves] to tell other men at home; and I want to take pictures of everything and show them to my friends" (1895:321). (Photograph by J. K. Hillers, Arizona Historical Society)

(one band had never been recognized by the federal government). As of 1957 (when the law went into effect), the 232 people who lived on 42,839 acres in Utah were no longer eligible for federal assistance programs. Part of the Truman and Eisenhower administrations' efforts to force American Indians to assimilate rapidly into mainstream society, this action by Congress was a devastating landmark in the history of the Southern Paiutes.

The consequences of termination were the same for the Nɨngwɨ as they were for 108 other American Indian tribes and bands. Their already low standard of living plummeted; disease increased due to the loss of free health care from the U.S. Public Health Service; unemployment increased through the loss of government employment, job training, and educational programs; and the loss of the collective land base resulted in the removal of a center for cultural identity, with a concomitant decline in customs and traditions.

In 1980, following a lengthy and expensive fight, tribal status was reaffirmed and the five bands—the Shivwits, Kanosh, Koosharem, Indian Peaks, and Cedar City bands—were reinstated as the Paiute Tribe of Utah. To demonstrate to the federal government that tribal status should be restored, the Southern Paiutes had to convince both Congress and the executive branch that there had been a decline in tribal members' health and welfare and that this would not have occurred if services had been available. The groups had to prove that they were among the poorest peoples in the area—ranking low in such sociological measures as total population, family income, employment, adequate housing, high school graduation rates, and the amount of land in productive use. In short, they had to establish that their standard of living had dropped since termination and that the group was no longer able to furnish its members with such basics as water and electricity.

Simultaneously the five bands had to show that there existed an ongoing, identifiable community of Indians who were members of the formerly recognized tribe or their descendants and that these individuals still lived in the vicinity of their former reserved lands, spoke the indigenous language, and continued to follow traditional customs and cultural practices. What was

"THE NAVAJO WEDDING BASKET"

Nancy J. Parezo

Modern contacts continue to influence Southern Paiute basketry. Many Navajo ceremonies are not complete without a ceremonial basket called by traders a "wedding basket." The banded, stepped design in red and black against a light background is an old Navajo basket pattern. Baskets of this style are often used by the Navajos in Blessingway ceremonies (for example, in weddings and the blessing of houses) and in curing ceremonies. The pattern is symbolic of the earth, mountains, sun, and clouds. While an essential item of ritual equipment, so many restrictions and taboos are placed on Navajo women while they make baskets that there are fewer and fewer basket weavers to meet the increasing demand for them.

Around 1890, San Juan Paiute and Southern Ute women, many of whom had intermarried with the Navajos, began to make baskets using the designs and techniques of the Navajos. These women were not constrained by the ritual rules that limited Navajo women's behavior during the construction period. In the twentieth century, the Navajos have relied almost solely on the Kaibab and San Juan Paiutes and the Southern Utes to fulfill this need. This intertribal trade is quite active and is a major source of income for some San Juan weavers. This means that the "Navajo wedding basket" for sale in shops was probably made by a San Juan Paiute woman. Navajo religious requirements, however, necessitate that Paiute weavers modify their traditional weaving techniques. Weavers produce baskets with new designs, a new shape, new selvedge treatment (false braid), and an interior work surface. Baskets have a three-rod bunched foundation coiled from right to left (clockwise) to conform to Navajo directional concepts.

In recent years, San Juan weavers have begun to make several new styles based on the Navajo wedding basket shapes. Included are Navajo *yei* designs and a number of traditional Navajo Spider Woman crosses and designs commonly associated with Hopi basketry. These are selling widely. Three generations of the Lehi and Whiskers family have worked hard to adapt Navajo and Paiute designs.

most difficult under the circumstances was the requirement that the Nɪngwɪ demonstrate that they continued to perform self-governing functions either through elected representatives or meetings of their general membership. Restoration was also highly political; they had to convince senators and members of Congress from the affected states (Arizona and Utah) to introduce and vote for the bill.

In 1990 the San Juan Paiutes of Arizona were also recognized by the federal government as an independent tribe (but without a land base) and not members of the Navajo Nation. Maintaining their language and cultural identity in the midst of the much larger Navajo population, with whom many had intermarried, had been difficult throughout the years, but the Nɪngwɪ succeeded. As of 1987, 190 individuals lived in household clusters at Willow Springs near Tuba City and at Navajo Mountain. The main farming area for San Juan Paiutes is in Paiute Canyon, although a few families also have gardens at other sites to the west. Since the early 1980s, San Juan Paiutes have been involved in legal cases to give them the right to intervene in the Hopi-Navajo land dispute and to obtain lands of their own. These latter cases are still in court, but the San Juan Paiutes are hopeful that they will soon have control of their homeland.

Today and the Future

The Nɪngwɪ have endured many hardships, but their relationship with the land and the intimate knowledge required for living on it has sustained them through the years. Today, the people live in small groups, wear western clothes, and reside in modern houses, but there are cultural boundaries that separate them from local Anglo American communities. Most visible are crafts (beading and basketry; see pls. 35–37) and dances. Less visible are the use of their native language and the maintenance of traditional beliefs and values. Band names are still used to designate Southern Paiute communities, visible labels of social divisions that are recognized by both Indians and local whites. Most Nɪngwɪ families go to one or more of the dances and powwows held during the year. At these events, acquaintanceships are renewed and news is gathered. These times of joy reinforce Southern Paiute identity. Families may also attend

Cry, or mourning, ceremonies, which are held three months to a year after a relative had died.

The most important social unit is still the family. The Nɨngwɨ feel that it should be the hearth of love and peace, and kin should act in solidarity, supporting one another's actions and being ready to help without judgment or criticism. Old persons are respected for their wisdom, long life, and support of the young during their nurturant years. Children are greatly loved and are felt to bring joy and comfort when life is hard. Conformity to this ethical system is viewed by the Southern Paiutes as a continuation of their cultural past and is used as a basis for membership in their ethnic community. The Nɨngwɨ find the prevailing individualism expressed in the Anglo American mass media contradictory to their beliefs and view it as self-serving and morally sterile (Knack 1980). Because of this, they try to limit the amount of interaction with non-Paiutes, a strategy that they used successfully in the past. Many people also continue to maintain older religious traditions while simultaneously subscribing to Anglo religions.

During the past twenty years, the San Juan and Kaibab Paiutes have made strong efforts to resolidify their culture by overcoming poverty and developing their lands. With funds from the 1970 settlement of a land claims case, the Kaibab Paiutes have developed an economic strategy that focuses on cattle ranching, farming, and tourism. Little hunting is possible, but regulated cattle raising has proven successful. The Kaibab Paiute Tribe leases their visitor center to the National Park Service and operates a nearby campground at Pipe Springs National Monument. They have modern tribal offices and well-developed housing facilities for tribal members.

The San Juan Paiutes face different problems. They struggle to maintain their language and tribal identity while living in the midst of the larger Navajo Nation. Their goals are to develop the economy of their own reservation and to secure self-determination. Today the Southern Paiutes are emerging from their position as one of Arizona's forgotten peoples. They are working to establish a strong position on this northern frontier. Through land-claims cases, federal recognition petitions, and economic development, they are taking control of their lands, their identities, and their future.

THE COMCÁAC (SERIS)
PEOPLE OF THE
DESERT AND SEA

Thomas E. Sheridan

In 1958, linguists Mary and Edward Moser recorded a shaman song sung by Antonio Herrera, a Seri living in the village of Desemboque on the coast of Sonora. The song describes the vision of a mythological giant who crossed the Gulf of California from Baja California to Tiburón Island off the Sonoran coast.

> Slender whirlwinds coming from the sky touch the land.
> Sounds of arrows
>   striking the ground,
>   roaring,
>   raising dust clouds.
> He shouts, warning of the day of danger.
> I stand on the peak of Red Mountain.
> He comes toward me
>   shouting.
> My heart is as stone.
> I shout, I declare it.

The song, like all the other Seri legends about giants, is important because it may shed light on the origin of the Comcáac, or Seris, themselves. The Seris are a small group of fishers, hunters, and gatherers who occupy Tiburón Island and the central coast of Sonora. Surrounded on one side by the Gulf of California and on the other by a desert of relentless aridity, the Seris have used both the desert and the sea to keep the rest of the world at bay. Unlike their neighbors, the Yaquis, Opatas, and O'odham, the Seris never farmed their dry domain. Instead, they relied entirely on wild plants and animals to survive. Further, even though most linguists place their language in the Hokan stock, the relationship of Seri to other Hokan languages in the region (Cocopa, Quechan, and Mohave, among others) is extremely remote. Linguistically and culturally, the Seris bear little resemblance to the Native Americans of Sonora. Where, then, did the Comcáac come from?

The answer may lie in the lore about giants. According to legend, a race of giants occupied both Baja California and coastal Sonora when the Seris first moved into the region. The Seris do not believe they are descended from the giants, but the giants clearly are associated with all that has become obsolete in modern Seri culture. Giants fashioned the characteristic Seri thin-walled "eggshell" pottery (see pl. 38) and made deep circular *metates* to grind mesquite pods and seeds. Giants wore stones and shells in their noses and ears. The Seris made such artifacts and practiced such customs in the past, and the fragments of "giant talk" that have survived are nothing more than archaic forms of the Seri language. So when the Seris sing about giants wading across the Gulf of California, they may be mythologizing their own migrations from the Baja peninsula.

They certainly had the technology and cultural orientation to do so. Seri bands inhabited both Tiburón and San Esteban Islands, paddling across the open sea in their canoes, called *balsas*, which were made of three bundles of *carrizo*, or native cane, bound together with mesquite roots. During the Spanish colonial period, Seri raiders occasionally attacked missions in Baja California, including Loreto, which lay far to the south. The Seris were a seafaring people accustomed to island-hopping across the gulf. To them the sea was an avenue, not a barrier.

The Seri way of life also bore more resemblance to the societies of Baja California, particularly the Cochimí of Baja's central desert, than it did to those of the Sonoran mainland. Both the Seris and the Cochimí occupied extremely arid terrain and practiced no agriculture. Both broke into small groups to forage for wild plants and animals, and both used balsas to fish and to transport themselves across the sea. Moreover, both spoke Hokan languages, although the relationship between the Seris and the Cochimí has not been explored in any detail. Sometime

Ramón Blanco poling a reed balsa with a turtle harpoon in Kino Bay in 1922. (Photograph by Edward H. Davis, Smithsonian Institution, National Museum of the American Indian)

in the past, then, ancestors of the Seris may have migrated from Baja to Tiburón Island and the Sonoran coast.

The Sea in Seri Culture

Regardless of where they came from, however, no other American Indian people in the Greater Southwest relied more on the ocean than the Seris. The gulf provided them with food, mobility, and refuge. When their enemies attacked them on the mainland, they fled to Tiburón or San Esteban. When Spanish or Mexican soldiers invaded the islands, they escaped to the coast. The presence of the ocean, with all its resources, made Seri territory a land of abundance rather than scarcity. It also allowed the Seris to remain outside the control of empires and nation-states until the twentieth century.

But crossing open water could be a terrifying experience. Even with the wooden boats powered by outboard motors that Seri fishermen use now, treacherous currents and gulf storms often strand Seri fishing parties on the islands for days at a time. Imagine facing those currents in a low-slung balsa propelled by

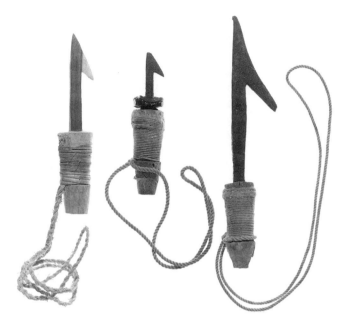

Seri harpoon points made (left to right) before 1941, before 1955, and before 1963. The point at far right was made by Jesús Morales. (Photograph by Helga Teiwes, Arizona State Museum)

a double-bladed paddle. An old Seri woman named María Antonia Colosio recalled the perils of one of those voyages she made as a child in the early 1900s:

> We were at *Cyajoj* [the south shore of Tiburón Island]. We were on two balsas that were tied together, side by side. . . . I didn't want to go. We were going to *Coftécöl* [San Esteban Island]. . . . I didn't want to get on, but my father caught me and put me on. After he caught me and put me on, he tied me behind a blind man who went along to paddle. Then I cried a lot, but he didn't pay any attention to me. That's how we went to Coftécöl. It was so dangerous when we almost entered the area called *Ixötáacoj* (Big Whirlpool). The sea just swirled and churned. The wind wasn't blowing but the water was choppy. It just churned. It was dangerous. The sea was going around. Everything just roared. The children and old women all cried. The old man Pozoli just said, "We'll land really soon." As we

were going to land, he sang to the shore. And it seemed we landed right away. The men paddled with all their strength, and we landed near the rocks.(Felger and Moser 1985:131–32)

The Seris did not use either nets or hooks until the advent of commercial fishing in the 1930s, but their balsas enabled them to spear a wide variety of fish, including mullet, snappers, groupers, and sea bass. They used a double-pronged harpoon with serrated points made from catclaw, which were bound to the shaft with mesquite-root twine. Seris also employed a single-pronged harpoon to spear fish from the shore. Their most important marine prey, however, were the five species of sea turtles found along the Pacific Coast, particularly the green sea turtle (*Chelonia mydas*). The Seris, in fact, identified eight different types of green sea turtles by name, distinguishing a number of subspecies that had probably died out by the mid twentieth century. Even though the Seris' technology was simple, their knowledge of marine life was both precise and profound.

Ethnologist W. J. McGee, who visited the Seris in 1894, estimated that sea turtle meat provided 25 percent of the Seri diet. The Seris also used the turtles' fat for cooking, their stomachs as water bags (see pl. 40), their leathery skin as sandals, and their carapaces as containers and building material. The peak hunting season occurred during the warmer months, when young turtles migrated north into the Gulf of California from nesting grounds in Michoacán and other southern coasts. Two of the twelve moons of the Seri year reflected the importance of this hunt: February, the "moon of few turtle hunters" (*cayaj zaac*), when the migrating turtles were just beginning to arrive, and March, the "moon of many turtle hunters" (*cayaj-aacoj*), when large schools of turtles fed in the shallow coastal waters. Nevertheless, Seris harpooned turtles throughout the year, including dormant turtles who had buried themselves in the sand at the edge of eelgrass meadows in the Infiernillo Channel between Tiburón Island and the mainland. Beginning in the 1930s, when turtle meat and soup began to be considered delicacies in northwestern Mexico, both Seri and Mexican fishermen harvested turtles for the commercial market. Five decades later, sea turtles had become so scarce—primarily because of overharvesting on the nesting beaches of Michoacán—that the Mexican government banned the hunting of sea turtles except by the Seris for their own consumption.

Because of their importance, green sea turtles figure prominently in at least one version of the Seri origin myth, which reveals Seri dependence on the sea as well. In the beginning, it says, there was only the ocean. Then Hant Caai—"he who made the land"—created marine and terrestrial animals and placed them upon a huge balsa. With the help of a green sea turtle, Hant Caai shaped the land and had Hant Quizim—"he who hardens the land"—give it solidity. The first living thing Hant Caai placed on the earth was a red elephant tree (*Bursera hindsiana*). Under that tree he placed a man, a woman, and a horse. Those first people were giants, who later died in a great flood and were transformed into boojum trees (*Fouquieria columnaris*), barrel cactus (*Ferocactus wislizenii*), and other desert plants. Before the flood, however, Hant Caai subjected the first man to a series of tests in order to discover his natural skills. When he told the man to mount a horse, the man promptly fell to the ground. When he told him to paddle a balsa, however, the man headed out to sea and harpooned a sea turtle. This proved that the first man was a good fisherman but was unable to work the land (adapted from Felger and Moser 1985:100).

The resources that the Seris drew from the Gulf of California were not limited to fish and sea turtles. They also hunted California sea lions, especially on San Esteban Island, and gathered mollusks such as clams, oysters, and octopus from tide pools and estuaries along the coast. Shell middens up and down the Sonoran shoreline attest to the enduring importance of shellfish in the Seri diet. Shells also served as tools, utensils, and ornaments (see pl. 41). The Seris undoubtedly participated in the prehistoric shell trade that integrated the Trincheras, Hohokam, and Casas Grandes cultures in a far-flung network of exchange.

But no other resource reveals the Seris' intimate understanding of the sea better than eelgrass (*Zostera marina*), a marine grass that grows submerged along coastlines around the world. Within historical Seri territory, eelgrass meadows are found just below the tide zone from Guaymas Bay to Desemboque. Eelgrass meadows are particularly abundant in the Infiernillo

Channel, where they attract large numbers of sea turtles, who feed on the long, willowy shoots. Consequently, Seris often hunted sea turtles in eelgrass meadows, claiming that the meat of the animals who fed there was particularly sweet.

The primary use of eelgrass, however, is as a grain consumed by the Seris themselves. During the winter, the plant produces seed-bearing shoots that ripen in April and early May. The shoots break off and float to the surface, and the Seris wade into the water and gather great masses of them. The shoots are then dried on the beach until the seeds can be thrashed, winnowed, and ground into flour by Seri women. For that reason, April is known as the "moon of the eelgrass harvest."

Seris are the only people in the world known to have utilized eelgrass seeds, or the seeds of any other sea grass, as a food staple. And they have been using it for a very long time. In the seventeenth century, Jesuit missionary Andrés Pérez de Ribas wrote that the Seris "gather the small seed of a grass, which grows under the water in the sea, which also serves them for bread" (Sheridan and Felger 1977:91). They may not have cultivated terrestrial grains, but they knew the plants of the Gulf of California well enough to extract a grain from the sea.

The Desert in Seri Culture

Seri knowledge of the desert was even more extensive. The Seris had names for about 400 native plants, of which 94 species, nearly 25 percent, served as food. Among the staples were the fruit and seeds of the four columnar cacti that grow along the Sonoran coast (saguaro, organ pipe, *pitaya agria*, and *cardón*). The Seris harvested the fruit of saguaro and cardón with saguaro-rib poles similar to those used by the Tohono O'odham. They also fermented the fruit of all four species to make wine, but consumption was not associated with ritual as it was among the Tohono O'odham. Both cardón and organ-pipe fruit were dried and preserved. The Seris stored cardón seeds as well, placing them in sealed *ollas* (pots) and hiding them in caves because they were high in vegetable oil and mixed well with other foods, like eelgrass seeds.

Ramona Casanova trimming the leaves of a century plant near Desemboque, Sonora. The caudex (basal portion) of the plant was baked on a bed of coals and eaten. (Photograph by Richard Felger)

Another major source of food was mesquite. The pods were toasted, pounded, and winnowed to make a flour consumed as *atole* (gruel) or kneaded into dough. The Seris also chewed and twisted mesquite roots to make cordage, fashioned harpoons from mesquite roots and hardwood, extracted black paint and basket dye from mesquite bark, and boiled mesquite leaves and bark to make medicinal tea. No other plant was as versatile or as important to Seri culture.

Various species of agave, or century plants, provided a welcome sweetness in the Seri diet. When the agaves began to send up their inflorescent stalks in late winter, the Seris cut them off at the ground. Then they trimmed away the sharp, stout leaves and baked the whitish hearts in rock-lined pits. The high concentration of carbohydrates in the hearts made them much sweeter than most other plant foods, so children eagerly awaited

their harvest. Hearts not eaten immediately were dried and preserved. The Seris also fermented both hearts and leaves to make
wine.

Other important food plants included a small shrub called
coap by the Seris and *mala mujer* (*Cnidoscolus palmeri*) by the
Mexicans. Each shrub produces several dozen tuberlike roots,
which the Seris gathered year-round by prying them from the
rocks and roasting them in hot ashes. They also collected the
fruit of jumping cholla and desert wolfberry, and the greens and
tiny seeds of ephemerals such as amaranth, chenopodium, and
wooly plantain. Women and children did most of the gathering,
while men fished or hunted mule deer, javelina, and bighorn
sheep.

Because of the range of resources they exploited, the principal
limiting factor on Seri society was water, not food. The coastal
desert of Sonora is one of the driest regions in North America.
No portion of Seri country averages more than ten inches of rain
a year, and some areas receive four inches or less. No rivers
rippled across the landscape. The only flowing water came from
three tiny streams: one at the south end of Tiburón Island, another in Nacapule Canyon north of Guaymas, and the third in
the Sierra Bacoachi northeast of Playa San Bartolo. All three
were less than a mile long. On the Sonoran coast there were
only five permanent sources of water between Puerto Libertad
and Bahía de Kino, a distance of about 130 miles. These were
springs that seeped into bedrock pools or wells that the Seris dug
in arroyos.

The Seris survived by moving back and forth between these
few permanent water sources and by drinking from temporary
water holes after heavy rains. They filled their thin eggshell
pottery vessels and carried them miles to camp. They trans-
ported water in the stomachs of sea turtles and mule deer, and
even hacked open barrel cacti and sucked the juicy but bitter
pulp during droughts. Knowledge of seasonal or emergency water sources was critical because the Seris had to roam across wide
areas to fish, hunt, and collect plants as they ripened. On Tiburón Island, for example, they gave names to forty-three different locations for water, but only twelve or thirteen places
yielded water year-round. Consequently, there must have been

an intricate coordination between the search for water and the search for food. One without the other meant privation or death.

The barrel cactus (*siml*), in fact, was considered one of the most powerful plants in the desert. It possessed strong Icor, an invisible spirit that controlled the life of each plant. According to the Seris, Icor caused the barrel cactus to form fog, which created clouds and rain. In the words of one shaman song collected by the Mosers,

> All of the barrel cactus that grow on this land
> Fog coming from them
> They all have life
> It is the sound of the land
> It is the sound of the land
>
> FELGER AND MOSER 1985:173

The Nature of the Ihízitim and Seri Society

No other factor determined the nature of Seri society more than the scarcity of water. Most springs and seeps provided barely enough drinking water for people and no water at all for irrigation or livestock. As a result, the Seris did not farm the land or live in permanent settlements. Instead, they broke into small groups and harvested the desert and sea. They survived by knowing their environment, not transforming it. They were generalists rather than specialists, utilizing a wide variety of plants and animals for food, medicine, shelter, and spiritual power. This adaptability made the Seris very difficult to conquer when Spanish soldiers and missionaries colonized the rest of Sonora in the seventeenth and eighteenth centuries.

We may never know the precise outlines of Seri society before contact with the Europeans. The Spaniards realized that the Seris were never a unified tribe with any form of tribal government. On the contrary, the Europeans recognized a number of geographical subdivisions among Seri-speaking people. The Tepocas occupied the northernmost reaches of Seri territory, including the Bacoachi drainage. The Salineros ranged east from Bahía de Kino to the San Miguel River. The Tiburones lived on Tiburón Island. Finally, the Guaymas and Upanguaymas roamed the coastal desert as far south as Guaymas Bay and the north bank of the Yaqui River, where they intermingled with the Yaquis and Lower Pimas. The Spaniards referred less frequently to groups such as the Tastioteños, Carrizos, and Bacoachis, but the designations seemed to refer to the places where Seris were living when the Spaniards encountered them rather than to separate groups.

It is extremely difficult to determine the nature of these subdivisions or whether they reflected distinctions made by the Seris themselves. During the 1950s, Seris told linguist Edward Moser that six bands had existed in the past (Moser 1963). The bands were autonomous and occupied discrete territories. Furthermore, at least three of the bands were divided into social units called *ihízitim*, which were probably related groups of families with strong economic and emotional attachments to certain parts of the Seri territory. Bands and ihízitim may have considered those areas their spiritual homelands and may have had established use-rights to their natural resources and sources of water. Their range was not restricted to those areas, however, nor did they necessarily fight to keep others out. The Seris had to be able to wander across large expanses of the desert in their search for food and water. During times of drought, they also had to be able to gather around the few sources of water that had not run dry. Competition for water may have caused some of the "ancient passions and wars" that existed between the Tepocas and Salineros when Spanish troops under Juan Bautista de Escalante entered their territory in 1700, but water sharing undoubtedly brought different Seri groups together on many occasions. The Seris had to be flexible about where and with whom they lived; otherwise, they would not have survived in their riverless domain. Such flexibility certainly characterizes hunter-gatherers such as the South African Bushmen and Australian Aborigines in other arid areas of the world.

Empire of Sand: The Jesuit Mission Program

Social and geographic fluidity became even more essential when the Europeans rode into the Seri universe. It is interesting to note that a horse figures so prominently in the version of the

Seri origin myth quoted earlier in this essay. Horses were domesticated in the Old World and introduced into North America by the Spaniards. Nevertheless, the horse became such an important symbol of the non-Seri way of life that the first man's failure to master it contrasted sharply with his ability to paddle a balsa on the open sea.

Spaniards first heard about the Seris when Alvar Núñez Cabeza de Vaca and his three companions wandered into Sonora in the early 1530s. "On the coast there is no maize," Cabeza de Vaca reported, and "they [the Seris] eat the powder of *bledo* [wild greens] and straw, and fish that they catch in the sea from balsas because they do not have canoes" (Felger and Moser 1985:10).

A century later, the first Jesuit missionary to the Yaqui Indians, Andrés Pérez de Ribas, in describing the "Heris," wrote:

> This nation is excessively wild, without pueblos, houses, or fields. They have no rivers or arroyos and drink from small ponds or charcos of water. They sustain themselves by hunting, although during the time of the maize harvest, they go with deer hides and salt, which they gather from the sea, to trade with other nations. Those closest to the sea also sustain themselves with fish. Within the same sea, on an island it is said, live others of the same nation, whose language is held to be extremely difficult. (Felger and Moser 1985:11)

All the early Spanish descriptions stressed how different the Seris were from the people trying to subdue them. They did not farm. They did not live in towns. Most American Indians seemed barbarous to the European intruders, but the Seris appeared almost incomprehensibly savage. In the words of the Jesuit Adam Gilg, who tried to missionize them,

> They roam like gypsies from one place to another (without staying anywhere long) in order so that they can gather the vegetation, fruits, herbs, and seeds which nature produces by herself without work; and for this purpose they prefer to visit those regions in which they find better food of that sort, according to the season of year. Anyway, they live like cattle, without God, without law, without faith, without Princes, and without houses. (DiPeso and Matson 1965:43)

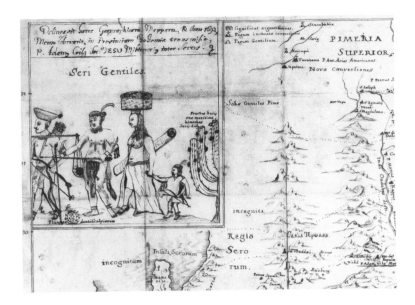

A 1692 map drawn by Padre Adam Gilg. Gilg was one of the first Jesuit missionaries to the Seris. (Smithsonian Institution, National Museum of Natural History)

Such culture shock is not surprising; few people were as different as the Seris and the Spaniards. The Spaniards were products of a rigidly hierarchical society dominated by church and state. The Seris lived without priests or political authorities except shamans and war chiefs. Nonetheless, Jesuit missionaries spent nearly seventy years striving to make the Seris over in their own image. Padre Juan Fernández founded a mission called Santa María del Pópulo in 1679. It was followed by Nuestra Señora de los Angeles in about 1700. Both were located on the San Miguel River below the Eudeve Indian missions of Opodepe and Cucurpe. Because there were no rivers in Seri territory, the missionaries led their neophytes to water. It was a policy known as *reducción*, the "reduction" of the Indians to civilized life. The Jesuits wanted to turn the Seris into farmers and town dwellers as well as Catholics. Missionization always involved much more than religious conversion.

Despite their endeavors, however, the missionaries met with only limited success. The height of the mission program was reached in the early 1730s, when Padre Nicolás Perera claimed that 700 Seris lived at Pópulo. Perera may have exaggerated the number for political reasons, but in any case most of the mission Seris soon returned to the desert or to Tiburón Island. The rest of the time the population of Pópulo and Los Angeles fluctuated between 100 and 200 individuals, perhaps 10 to 20 percent of the total number of Seris. While some of the Seris may have been genuinely attracted to the new religion and the new way of life, others treated the missions as little more than another stop on the seasonal round. When times grew lean, they joined the missions. When cactus fruits were ripe or sea turtles were feeding in the eelgrass meadows, they drifted back to the coast. Life in the desert had turned the Seris into consummate opportunists. Life in the missions was never strong enough to change them.

As early as the 1640s, the Seris also began to kill Spanish cattle, which must have seemed like slow, dumb deer to the desert hunters. They never developed raiding economies as pervasive as the Apaches or Comanches, but rustling became a chronic source of tension between the Seris and their Spanish neighbors. In 1725, for example, Salvador de la Huerta of Opodepe hanged three Seris for stealing his cattle. Sixty kinsmen of the rustlers retaliated by setting de la Huerta's home on fire while he was hosting a fiesta. Twenty-one men, women, and children were clubbed to death or shot with arrows as they tried to escape the flames. Similar episodes of provocation and revenge embittered Seri-Spanish relations throughout the colonial period.

But a complete breakdown did not occur until the mid eighteenth century, when Spanish officials transferred the *presidio* (military garrison) of Pitic to San Miguel de Horcasitas, on the San Miguel River between Pópulo and Los Angeles. Even though missionaries warned of the consequences, presidial soldiers were soon forcing the Seris to grow their crops and raise their stock. In the words of Padre Tomás Miranda, the Seris' last Jesuit defender, "The people work every day from sunup to sundown, poorly fed and badly mistreated without exception, including pregnant women" (Sheridan 1979:324).

Such abuses drove most mission Seris back to the desert, where they joined forces with their gentile brethren to wreak havoc on central Sonora. To end the bloodshed, the viceroy of

A Seri family at Rancho Costa Rica in 1894. Face painting was a popular adornment for Seri women. (Photograph by William Dinwiddie, Smithsonian Institution, National Museum of Natural History)

New Spain issued a general pardon, and eighty families trickled back to the missions. But Spanish officials in Sonora placed little faith in friendly persuasion. Rather than resettling the Seris in the missions, they proposed to deport them to the Caribbean. "Once secured in a boat," wrote José Rafael Rodríguez Gallardo, the architect of this final solution to the Seri problem, "they will only be able to seek their freedom in their own shipwreck and ruin" (Sheridan 1979:326).

By then, most Jesuits concurred with Rodríguez Gallardo's assessment. In a remarkably cynical document, seven missionaries joined the governor of Sinaloa-Sonora in declaring, "The extraction of the Seri called Populeños, Salineros, Tepocas, and Tiburones is not only necessary but requisite and urgent. Even though they are distinguished by names which have been taken from their place of habitation, all this multitude is the same nation with the same nature, equal in propensities and inclination. They are an arrogant people devoted to life in the wild" (Sheridan 1979:327).

The result was two decades of fire and bloodshed. Governor Diego Ortiz Parrilla assembled a force of more than five hundred presidial soldiers, Spanish militia, and Upper Pima auxiliaries to invade Tiburón Island. The governor proclaimed the invasion a great success, but only thirteen Seri men were killed and twenty-eight men, women, and children taken prisoner. The rest broke up into small groups and fled to the Sonoran mainland, where they took refuge in mountain strongholds like Cerro Prieto (modern Sierra Libre) northeast of Guaymas or hid in the mangrove estuaries dotting the Sonoran coast. There they became a nearly fatal cyst in the Sonoran body politic, joining forces with both Upper and Lower Pima rebels to seek vengeance on their Spanish foes.

Mobility and Attrition

The strength of the Seris lay in their mobility. They raided Spanish settlements and then disappeared into the desert, where only they knew where the water was. Time after time the Spaniards mounted major campaigns against them, and time after time the Seris slipped through the Spanish net. The largest was the expedition of Colonel Domingo Elizondo, which consisted of more than 1,100 troops, including 700 professional soldiers. In 1768, Elizondo established his naval base at Guaymas and his barracks at Pitic, which later developed into the Sonoran state capital of Hermosillo. He invaded the rugged canyons and impenetrable thorn forests of Cerro Prieto on two occasions, hoping to crush the Seris and their Pima confederates once and for all. Both times the Seris eluded him. The flexibility of their society and their knowledge of the desert and sea made them superb guerrilla warriors.

Nevertheless, the sheer size and persistence of the Elizondo expedition wore many Seris down. After the last invasion of Cerro Prieto failed in 1769, Elizondo broke his forces into small detachments and fought a guerrilla war himself. Forays led by seasoned frontier commanders like Juan Bautista de Anza scoured the coast, harassing and killing Seris in mountain ranges, mesquite thickets, and mangrove swamps. Some of the Comcáac fled to Tiburón Island. Others surrendered and asked for clemency. Fifty-nine families were given land at Pitic, where they formed a community that became known as Villa de los Seris. Crisanto, one of the foremost Seri leaders, even journeyed to Mexico City to receive a scroll from the viceroy bearing a general pardon.

The impact of the Elizondo campaign on Seri society must have been immense. More than twenty years of constant warfare decimated the Seri population and weakened boundaries among different Seri subdivisions. As people were killed or taken prisoner, survivors of one group merged with survivors of another. Subdivisions like the Guaymas and Upanguaymas disappeared. The Tiburón Seris must have taken in many Tepoca and Salinero refugees. Moreover, Seri territory shrank as Guaymas developed into a major port and Pitic into one of Sonora's largest towns. In the north, the gold placers of Cieneguilla, southwest of Altar, attracted thousands of Spanish prospectors in the 1770s. By the end of the eighteenth century, the desert was no longer such an effective barrier for the Comcáac.

Those patterns of attrition continued during the nineteenth century. After Mexico won its independence from Spain in 1821, Guaymas became one of the most important ports on the new republic's west coast. That in turn stimulated the growth of Pitic, which changed its name to Hermosillo. Mule trains and caravans of freight wagons plodded across the desert between the two commercial centers, attracting Seri raiders. Numerous legends sprang up around one such raid in 1850, when a Seri named Coyote Iguana abducted a young Mexican girl named Lola Casanova, who bore him at least one son. In the second half of the nineteenth century, however, a rancher named Pascual Encinas established his headquarters at San Francisco de Costa Rica on the margins of Seri territory. At first Encinas tried to employ Seris as laborers, but as more and more cattle spread through the desert, Seri livestock raiding increased. That led Encinas to turn his cowboys loose against the Indians, who were hunted down like predators. The old rancher, who was inter-

SERI IRONWOOD CARVING

Thomas E. Sheridan

For thousands of people across North America and Europe, the signature of the Seri Indians is their smooth, dark-brown animal figurines. Carved from wood of the ironwood tree (*Olneya tesota*), which grows along arroyos at lower elevations of the Sonoran Desert, the best of these sculptures burn with a rich, heartwood luster that captures the essence of sea turtles, pelicans, bighorn sheep, and other creatures of the desert and sea. You can find them in Indian art stores and in the homes of collectors around the world. They have become one of the principal industries of the Comcáac, a major source of cash to buy pickups, clothes, food, and fiberglass boats—all the commodities of late-twentieth-century life that the Seris need.

But ironwood sculpting is not some ancient Seri craft passed down from generation to generation. In the past, Seris fashioned dolls and fetishes from wood, but it was usually soft wood like the red elephant tree (*Bursera hindsiana*). Ironwood, on the other hand, is one of the two densest woods in North America (the other is Florida leadwood). Stone tools were no match for its heavy grain.

By the early 1960s, however, the Seris had

Nacho Burnett and his wife, Mercedes Diaz, carving ironwood at Punta Chueca in 1982. (Photograph by David Burckhalter)

obtained hatchets, metal files, saws, and sandpaper, so one enterprising individual—José Astorga—began applying those modern tools to the hard tree. The Seris were no strangers to ironwood. They already carved harpoon shafts from its wood, ate its ripe pods, and even made a mash from its seeds to calm the wind-rippled water of estuaries while fishing. But Astorga was the first to coax the shapes of mammals and birds from its burnished wood.

There are numerous versions of why he did so. Astorga himself claimed that he had a vision while fasting in a cave not far from Desemboque, where he lives. After several days without food or sleep, a batlike spirit appeared to him and told him to sculpt ironwood figures to save the Seris. The spirit even showed him some of the forms to carve (Yetman 1988). Individual vision quests are an important element of Comcáac spirituality, so Astorga's account fits into a time-honored Seri pattern. Others claim that North American friends of Astorga encouraged him to work with the wood. Whatever its origin, however, ironwood carving has developed into an art form that is also one of the Seris' primary links with the outside world. If imitation is the sincerest form of flattery, hundreds of Mexican craftsmen pay tribute to the Seris every day, grinding out inferior figurines in mechanized shops. Cute little owls with big eyes now flood the market, crowding out the simple, graceful creations of the Comcáac in many southwestern shops.

The insatiable demand for ironwood has also taken its toll on regional populations of the tree. Ironwood is not endangered yet, but Seri carvers now have to pay more or drive farther to find trees big enough to provide them with the wood they need. Competition from Mexican carvers is just one of the many pressures the Comcáac feel from the hungry, growing nation that surrounds them.

SERI SANTOS

Diane Dittemore

In 1826, Lieutenant Robert W. H. Hardy, a British naval officer searching for pearl beds in the Gulf of California, visited the Seri camp of Tecomate on the north shore of Tiburón Island. There he reported seeing "a wooden figure with a carved hat, and others of different shapes and sizes" (Hardy 1829:294–95). More than a century later, anthropologist Gwyneth Harrington Xavier collected several hundred Seri objects for the Arizona State Museum. Among these were *santos,* carved talismans made of red elephant tree wood (*Bursera hindsiana*) and painted with red and blue commercial pigments (see pl. 43). "They may be seen worn around the neck," she wrote, "tied to the house doorway, and occasionally tied at regular intervals all around the inside structure of the house" (Xavier 1941:34).

Although any Seri could make a santo, those made by medicine men had the greatest power. Respected medicine man Santo Blanco told Xavier that when santos were tied to a house, "it was to prevent sickness, evil or death from entering that house. Worn on the person, they were to cure a present illness. When a person was seriously sick, the figurines were taken to a cave where the spirits live . . . and sometimes the medicine man sang to the santos there for a cure. The medicine man makes these santos for pay, and at times fasts and prays and sings for curing" (Xavier 1941:34).

The santos Xavier collected came in many shapes and sizes. Men in hats or those with crosses were considered *católico,* meaning simply spiritual, while naturalistic representations such as the moon and birds were "pura Seri." A pair of charms tied together with a string was called a household santo and was hung in the doorway of the house to protect the people who lived there. One was the figure of a man standing on the crescent moon with a cross on his chest. "This is a thing for the man," Xavier recorded. "This santo stands upon the new moon. This santo can look both ways, he has two faces, he looks both ways, below the earth and above the sky. He is called 'Ah-kot-meh-ke,' and he comes at night to protect against death." The other figure was a manta ray. "This animal of the sea is a woman's thing. In Seri it is called 'kai-*nech*-in' [giant ray]. . . . It is for protection against death and disaster" (Xavier 1941:42).

Not all carvings in the shape of birds, snakes, and fish were classified as santos. Blue birds were *hano,* or messenger hummingbirds. According to the Seris whom Xavier consulted, "These birds appear to the medicine man when he has had no food or water for four days, sometimes eight days. Then the medicine man becomes the same as drunk, 'lo mismo borracho', many things appear, many people, talking. Great boats come in near, aeroplanes, and these birds come, bring messages, for the medicine man only. They tell of sickness, or whatever is happening far off" (Xavier 1941:44).

Absent from Xavier's collection is a type of santo most commonly encountered by tourists and traders since the 1950s. This is a man wearing a hat and standing within a door frame. It symbolizes the little people who inhabit the caves where medicine men embark on their vision quests. A medicine man named Nacho Morales apparently was the first to make these figurines. Within a decade after Xavier made her collection in 1940 and 1941, Seri use of the santos had nearly disappeared. Instead, the figurines were being carved for sale to the growing number of outsiders visiting Seri territory.

Don Pascual Encinas (seated center) with W. J. McGee (seated left) in 1894.
Encinas's cowboys killed hundreds of Seris during the mid nineteenth century.
(Photograph by William Dinwiddie, Smithsonian Institution, National Museum of Natural
History)

viewed by W. J. McGee of the Bureau of American Ethnology in
1895, estimated that about half of the Seris died during the
twelve-year "Encinas War."

The response of the survivors was to retreat deeper into the
desert or to take refuge on Tiburón Island. By the early 1900s,
only about 200 Seris remained. They were reviled as blood-
thirsty savages by the Mexicans and as cannibals by North
American journalists, a slur that had no basis in fact. Even
McGee, a geologist and anthropologist who led expeditions to
Seri country in 1894 and 1895, considered the Seris barely human. In his own pseudo-scientific words,

> Certain characteristics of the tribe strongly suggest lowly
> condition, i.e., a condition approaching that of lower animals,
> especially of carnivorous type; among these are the specific
> color, the centripetally developed body, the tardy adolescence,
> the defective tool-sense, the distinctive food habits (especially
> the consumption of raw offal and carrion), the independence

Seri women waiting by a tourist's car with crafts for sale in Punta Chueca, Sonora, in 1971. (Photograph by David Burckhalter)

of fixed habitations, and the extreme alternations between the rage of the chase and war and the quiescence of sluggish repose. (McGee 1898:156)

McGee was seeing the Seris at their nadir, after they had been nearly exterminated by a century and a half of brutal warfare. Their numbers had declined, and many of them survived only by begging at ranches like Costa Rica or at a clubhouse established at Bahía Kino in 1930 by U.S. sportsmen evading Prohibition. In 1941, when archaeologist Julian Hayden visited Tiburón Island, there were only 160 Seris left. Hayden and others concluded that the Comcáac were on the verge of extinction.

But the resiliency of Seri culture soon reasserted itself. Beginning in the 1920s, Mexican fishermen settled at Bahía Kino, on the southern margins of Seri territory. Most Seris moved to a camp at the edge of the village, where they learned how to make and use wooden plank boats (*pangas*), fishhooks, and nets. During the 1930s and 1940s, that process was accelerated by the demand for shark livers, a source of vitamins A and D, and shark

Lola Blanco with necklaces for sale outside Oscar Topete's store in Desemboque, Sonora, in 1971. (Photograph by David Burckhalter)

fins and the swim bladders of totoaba (*Totoaba macdonaldi*, the largest member of the croaker family), which were considered a delicacy by the large Asian population of California.

This series of boom-and-bust fisheries in the Gulf of California (the shark liver market collapsed after vitamins A and D were synthesized in 1941, and totoabas are now an endangered species because of overfishing) transformed the Seris from subsistence to commercial fishermen. For the first time in their history, they became integrated into the market economy of the region. Mexican traders bought their fish and turtles. In return, the Seris bought sugar, flour, and coffee, as well as lumber, putty, nails, tools, and, later, outboard motors and gasoline. In 1938 the Mexican government even organized a fishing cooperative among them at their old camp of Haxöl Ihoom (Place of the Clams), which became the village of Desemboque. In one sense, this made the Seris more dependent on Mexican middlemen and the Mexican state. In another sense, however, it deepened their dependence on the ocean and gave them greater mobility. Pangas powered by outboard motors carried them farther and faster than could reed balsas. By the 1950s the Seri population had climbed above two hundred again, and they no longer needed to beg.

Then, in the 1960s, hundreds of North American tourists began wintering in Bahía Kino. With characteristic innovative flair, the Seris took advantage of the influx by creating an entirely new art form in Seri culture, the carving of ironwood figurines. José Astorga fashioned the first carving in 1961, a simple paperweight that he gave to his friend, the legendary bush pilot Alexander "Ike" Russell. Soon Astorga and others were sculpting hardwood figures of sea turtles, whales, pelicans, quail, and other animals of the desert and sea. The bold stylization and simple, flowing lines of these carvings quickly captured the imagination of collectors and tourists. Within a decade, more than half the Seri adults were making ironwood figures, and the sale of the sculptures brought in nearly as much income as fishing.

Today, there are more than five hundred Seris, and the population continues to grow. Much traditional knowledge has been lost, but the Seris continue to shape their own destiny, weaving beautiful baskets from limberbush (pl. 42), fishing, carving ironwood and wooden figurines called *santos* (pl. 43), and gathering desert plants. In the words of Richard Felger and Mary B. Moser, "the Seri were meeting the challenges of the modern world in much the same way as they reacted to earlier Spanish-Mexican influences—by taking advantage of new resources, new markets, and new opportunities without radically altering their loose and flexible way of life" (1985:19). That is the essence of Seri culture: independence, adaptability, and fluidity. Even though the marine resources of the Gulf of California have been decimated by commercial fishing, and the Seris' reliance on wild plant and animal foods has declined, they still make their living from the desert and sea. The fundamental relationships have not changed.

THE COLORADO
RIVER YUMANS
RELATIONS ON THE RIVER

Bruce E. Hilpert

On January 24, 1855, leaders of the Cocopa and Quechan tribes met with a group of u.s. Army officers aboard a steamboat in the Colorado River to arrange the terms of a truce, but the meeting was not meant to bring peace between the soldiers and the local Yuman-speaking Indians. Rather, the army officers were assisting in forging an agreement that would put an end, at least temporarily, to the sporadic warfare that had continued between the two Yuman peoples for generations.

As the afternoon wore on, Pasqual, Caballo-en-Pelo, and Vincente made speeches on behalf of the Quechan people. The leaders of the Cocopas—José, Jepita, Coyote, and Colorado—then presented their position. Conditions of peace were finally agreed to, and a treaty was signed by all the parties.

Lieutenant Nathaniel Michler, who witnessed the day's events, recorded that the Quechan had been especially anxious for the peace. He noted that they "had often sent down their women to propose one. Owing to the number of intermarriages among the tribes, this is the usual mode of proceeding" (Emory 1857:108). In his observations, the young officer made an important contribution to the written history of the Yuman-speaking peoples of the Colorado River region. He wrote about how these close neighbors regularly interacted: forming political alliances, intermarrying, and, despite their shared language, culture, and kinship, conducting warfare.

But Lieutenant Michler was not just an observer of these relations. As a representative of the u.s. government, he played a part in bringing far-reaching, irrevocable, and devastating changes to the peoples of the Colorado River. Beginning in the late 1840s, thousands of Anglo Americans and Mexicans passed through Yuman territory on their way to the California gold-fields. They became part of the multi-ethnic dynamics of the region, which was shaping the cultures of the Yuman-speaking peoples living there. Fifty years later as anthropologists began to document the cultures of the Cocopas, Mohaves, Quechan, and Maricopas, many aspects of the traditional way of life of the river peoples had already vanished. Later scholars turned to a variety of resources to help shed light on the cultural history of the Yuman-speaking peoples: detailed accounts of creation stories passed down through countless generations; historical documents left by Spanish priests; military reports prepared by u.s. soldiers; the often-fanciful accounts of travelers and settlers; and the memories of tribal elders.

Mastamho: Mohave Creator's Son

The creation stories of the Yuman speakers of the Colorado River contain detailed accounts of prehistoric life that range far beyond the military practices of the people. Through dreams, individuals were able to visit that time of creation eons ago when all people, plants, and animals came into being. These dream visits produced extremely detailed accounts of creation events that provided instructions on every aspect of life from farming, pottery making, and sexual practices to relations with neighboring tribes.

Each of the present-day Yuman-speaking tribes along the Colorado and Gila Rivers—the Cocopas, Quechan, Mohaves, and Maricopas—have similar accounts of the creation of the peoples of the region. In 1903 a Mohave elder named Chooksa homar provided ethnographer Alfred Kroeber with his own vision of the genesis, an account that took more than twelve hours to tell.

Chooksa homar's story begins with the death of the Creator, Matavilya. Matavilya had previously used his leg to stir the great darkness that prevailed—stirring and compressing the darkness to create the earth and the sky. Before his work was done, his daughter, Frog Woman, bewitched him and he fell ill and died. His young son, Mastamho, had to finish his work.

DREAMING FOR POWER

Bruce E. Hilpert

The most important feature of traditional Yuman spiritual beliefs was the special role of dreams. These nocturnal visions pervaded every aspect of daily life, from healing to warfare. A Maricopa elder, Xomace'kupa'ra, told anthropologist Leslie Spier: "Everyone who is prosperous or successful must have dreamed of something. It is not because he is a good worker that he is prosperous, but because he dreamed" (1933:236).

Through dreams, an individual received visits from powerful spirits that provided talents for singing, leadership, oration, or attraction to the opposite sex. For example, if the spirit of Buzzard chose a person to receive his power, he would begin to visit the dreamer while still a child. In the dreams, Buzzard would take the dreamer soaring over the desert, naming the mountains as he flew back home to Avikwame, the place of creation. There he would teach the youngster the songs that conveyed the power to heal the wounds from Yavapai arrows and to dream the location of the enemy during war.

The spirit would make repeated visits over the years until the lengthy songs were learned properly. To speak of these dreams could cause a loss of power or even death. In any case, speaking of the dreams was unnecessary since others would know of them when the power became apparent.

The spirits of a multitude of animals and natural phenomena such as stars and lightning conveyed very specific powers. Jackrabbit could cure blindness; Cricket cured the bewitchment caused by dreaming of the stars of Orion's belt; Frog gave the power to bewitch, to handle live coals, or sing for rain.

Dreams had other roles as well. Dreaming of dead relatives could lead to the loss of one's soul and resulting fainting spells. A married couple learned of the conception of a child through the man's dreams. A person could become lucky or unlucky by having dreams that were generally "good" or "bad." Bad dreams were especially common in the early evening and before dawn, so people's sleeping patterns were adjusted accordingly.

Today, the pervasive dreaming behavior and the elaborate song cycles have largely vanished among the Quechan, Mohaves, Cocopas, and Maricopas, yet dreams still hold a special place in the hearts of the people.

Mastamho told the Mohave people: "I will provide everything. The sun and the night have not yet been made, but I will make them; and I will tell you what to eat. Then you will know how to live" (Kroeber 1948:53).

One of Mastamho's first acts was to create the sacred mountain Avikwame (Spirit Mountain; Newberry Mountain in southern Nevada), recognized by all the Yuman peoples. Mastamho plunged a stick into the earth, causing water, fish, ducks, and a boat to spew forth. The water became the Colorado River, and Mastamho floated downstream, rocking the boat to form the broad river valleys that would become the fertile farmlands of the Mohaves and the Quechan. Mastamho then made plant foods for the hunting-and-gathering groups—the Chemehuevis, Hualapais, and Yavapais—and taught them to speak their languages. In conclusion, he said:

> You can go, you Hualapai, and scatter in the mountains there. You need not go into one place. You can go all about, for I have made springs everywhere. You can live in one spot, and when you want to live in another you can do so. You Chemehuevi can do the same, and you Yavapai too. But I will do differently for the Mohave. They will have everything along the river: whatever grows there will be theirs. It is well. (Kroeber 1948:56)

Mastamho then taught the people how to build a shade ramada and a winter house of logs and earth. He invited the Mohaves to enter the house, instructing the Chemehuevis, Hualapais, and Yavapais to sit outside on the west, northeast, and southeast to show the respective geographic locations of these peoples to the Mohaves. Later, Mastamho segregated the people inside the house, thus creating the Quechan and Kamias (California Yumas) from the Mohaves, indicating the self-perceived cultural similarities of the River, Delta, and California Yumans.

Using Scaup Duck and Frog as surrogates, Mastamho showed the Mohaves how to plant their crops in the floodplain of the Colorado River after the spring floods had deposited a rich layer of silt on the bottomlands. Mastamho himself taught them the names of the melons, corn, wheat, and beans that they were to grow. He also gave them *metates* and pottery vessels to prepare the foods.

After providing for their physical needs, Mastamho taught the Mohaves how to obtain power through dreaming. If a person dreams of Mastamho's journey, he or she will become a great runner. If a man dreams of a hawk, he will become a great warrior. Mastamho then taught the Mohaves to fight, telling them, "I want you to rush and seize and kill and fight and take slaves" (Kroeber 1948:61). He showed them how to make a war club and instructed the Mohaves to practice on a man standing outside the house (a non-Mohave). With his work done, Mastamho transformed himself into a bald eagle and flew south to the sea, forgetting all the knowledge that he had taught the people.

The Mohave creation story related by Chooksa homar is a very powerful document, not only because of its sophisticated literary devices and symbolism but also because of its extremely detailed account of almost every aspect of Mohave culture. It provides young listeners with instructive definitions of all the things they would need to know to function in their culture— not just ways of making a living but also values, beliefs, and attitudes toward neighboring groups.

Tribal Traditions and Spanish Contact

The earliest historical accounts of the Mohaves, Quechan, Cocopas, and Maricopas present a picture that mirrors the way of life taught by Mastamho. In 1540 the Spanish explorer Hernando de Alarcón led an expedition in two small boats up the Colorado River as far as present-day Yuma. He noted the facial tattoos "covering their faces almost entirely" and the practice of having "their ears pierced with many holes in which they place beads and shells" (quoted in Forbes 1965:101). In the 150 miles of the river traversed by Alarcón, he met seven or eight different groups of Indians that either lived on the river or were visiting to fish and gather mesquite beans. Juan de Oñate, another Spanish explorer, traveled farther upstream sixty-four years later. He found that "adjoining this nation [the Mohaves], on this same

river is another called Bahacecha. They speak a somewhat different language, although as the difference is small, it is not a hin-
drance to commerce and communication among them, for it is as
if they were all of one language" (quoted in Forbes 1965:103).

These Spanish accounts provide tantalizing but sometimes
confusing evidence about the groups living along the Colorado
three hundred years ago. Ethnohistorians have spent decades
sifting through an array of different names and imprecise geographical locations to identify and better understand the peoples
of the Colorado. Their research indicates that there were probably eleven groups living on the river in the 1600s who spoke
languages of the Yuman family—either mutually intelligible
dialects or closely related languages that could be partially understood.

Linguists class these Yuman speakers into three branches: the
Upland Yumans, including the Hualapais, Havasupais, and Yavapais, who occupy the plateau region south of the Colorado River
where it flows westward through northern Arizona; the Mohaves, Quechan, and Maricopas (including the merged Halchidhomas and Kavelchadom), known as the River Branch, who
lived along the lower Colorado and Gila Rivers; and the Cocopas,
who, along with the now-vanished Halyikwamais and Kahwan,
formed the Delta Branch and lived near the mouth of the river
below the present city of Yuma. Other Yuman speakers, including the Kamia and Kiliwa, lived to the west in southern California and in Baja California.

The groups that are no longer distinct today merged with
other peoples as their populations were decimated due to disease
and warfare during the nineteenth century. When visited by the
Franciscan missionary Francisco Garcés in 1775, the Halchidhomas were a thriving group with a population of more than
2,000 people. They were sandwiched between their traditional
enemies, the Quechan and Mohaves, however, and warfare and
slave raiding took its toll during the ensuing decades. In the late
1820s, the few remaining Halchidhomas fled the river to settle
near Magdalena, Sonora. An epidemic ravaged the tribe while it
was in Mexico, and in 1838 the survivors headed north to live
with their friends, the Maricopas. Today their descendants consider themselves Maricopas but are still aware of their ethnic

heritage. By 1870 the Kavelchadom, Kahwan, and Halyikwamais had also ceased to exist as autonomous groups.

This merging of one group of Yuman speakers with a larger,
dominant society is not a recent phenomenon. Evidence for its
antiquity can be found in the *cimú'l,* or nonlocalized and exogamous kinship groups of the Quechan, called sibs by anthropologists. There were twenty-three of these patrilineal descent
groups in historical times, each having a descriptive name and a
distinct account of its origin. Each of these kinship groups was
ranked. The sib with the highest status among the Quechan was
known as the *xavca'ts Kwatchán,* or Quechan. According to ethnohistorian Jack Forbes, the names and lore of at least nine of the
Quechan groups indicate that they derived from non-Quechan
groups such as the Mohaves, Kamias, or Maricopas. As evidence,
he cites the fact that one of the cim'l, the *liots Kwestamuts,* was
considered to be very different from other Quechan because in
the old days its members did not cremate their dead, a practice
used by every other Yuman-speaking group along the Colorado.

It seems quite likely that in prehistoric times small groups of
people may have sought refuge with or married into the Quechan, pledging their allegiance to the larger tribe. Given the
close proximity of so many groups with similar languages and
cultures, and the fact that the sibs were exogamous (a person
could not marry within the group), it is not surprising that there
was a great deal of intermarriage. Spanish priests and soldiers
took note of this on several occasions. Indeed, Pablo Coclum, a
Quechan *kwoxot,* or civil chief, was actually a Kamia by birth. As
Lieutenant Michler observed, these kinship ties often prevailed
over political concerns and helped end frictions among the different groups.

Whether new peoples came as refugees or because of marriage, it was essential that they acknowledge membership in
their new tribe and the authority of its leaders. A distinctive
feature of many Yuman societies was true tribal identity and the
existence of chiefs whose authority went beyond that of the
consensual leadership found among many other peoples of the
Southwest.

As Hernando de Alarcón noted upon meeting the Halyikwamais in 1540, "all of the people obeyed a single head" (quoted

in Forbes 1965:86). The smaller bands had their own leaders, and these men counseled the chief, but the kwoxot was vested with the authority to make decisions and some limited ability to enforce them. This system was most pronounced among the Mohaves. Ethnohistorians Clifton Kroeber and Bernard Fontana describe their political organization as a system of central government with a chief, a subordinate council, and constituent geographical groups that each member represented (Kroeber and Fontana 1986:102).

Among the Quechan, band leaders counseled the kwoxot and acknowledged his authority but did not constitute a formal council. While anthropologist Leslie Spier reports that the Maricopas had hereditary chiefs, among the Quechan their selection was through consensus of the people and their band leaders (Spier 1933). As in a parliamentary system, a new leader was selected whenever a previous leader's base of confidence and authority had eroded.

Such a system of centralized decision-making undoubtedly made foreign affairs more efficient in a region where contacts with neighboring groups were so frequent and "boundaries" were poorly defined. To an extent, peoples along the Colorado shared territories and the resources within them. This is not to say that outsiders could move onto the river and begin farming on land claimed by local residents. Nonetheless, it was common for peoples from the southern California deserts and mountains to make seasonal camps among the farms of the Cocopas. The Mohaves also allowed the Chemehuevis to live along the river on land that was undisputedly Mohave.

The Yuman peoples had frequent contacts with more distant groups as well. Despite their sedentary lifestyle, they were avid travelers. In 1774, Juan Bautista de Anza reported that the Halchidhomas journeyed from their home above the junction of the Gila and Colorado Rivers "as far as the coast tribes of California, but also that they make this journey in four days," an average of fifty miles per day (Forbes 1965:62). This pace suggests that the sojourner was running, not walking. A daily average of fifty miles would overwhelm a modern marathoner; it would be just another day's run for a Mohave who had gained power by dreaming of Mastamho's journey.

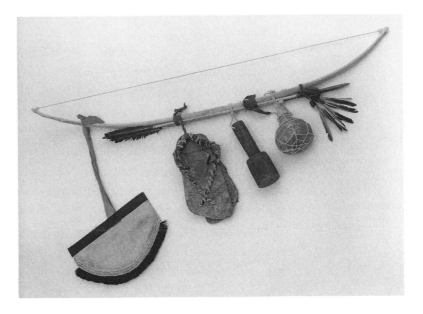

A Mohave traveler's kit, made by Romeo Burton in 1962. Such kits would have
been carried on long-distance trips. It is complete except for a bag of mesquite
beans. (Photograph by Helga Teiwes, Arizona State Museum)

Treks of up to a hundred miles were frequently made to visit
friends and relatives, or perhaps just to see new sights. To obtain
luxury goods such as seashells or fine textiles, traders commonly
embarked on journeys of more than four hundred miles. Archaeologists know that since prehistoric times California Indians supplied shells as raw materials for jewelry makers in
Arizona. Shell beads, abalone pendants, and carved shell brace-
lets are common finds in Arizona sites dating from the beginning of the Christian era.

In historical times the peoples of the Colorado occupied a
strategic position in a regional trade network. The Spanish priest
Pedro Font noted in 1769 that the Chumash Indians of California traded shells to the Mohaves and also possessed cotton blankets like those made by the Gila Pimas. At about the same time,
Garcés observed that the Tohono O'odham of southeastern Arizona had a "great abundance" of Hopi blankets that they had
received in trade with the Yumans of the Colorado (Forbes 1965:
148, 158). Not only were the Yumans centrally located "middlemen" between the population centers to the west, north, and

east, but they could also block passage across the Colorado River if they wished.

It appears that traders of different tribes—Quechan, Cocopas, Halchidhomas, and Mohaves—had trading partners in villages as far away as the Zuni pueblos in eastern New Mexico. These contacts were selective, since certain Yuman tribes were on friendlier terms with some distant groups than others. This friendship was due to a regional system of alliances that spread beyond the Yuman speakers. Thus the Halchidhomas' hostile relationship with the Quechan may have caused them to avoid the Yavapais but trade actively with the nearby Hualapais.

These tribal alliances arose from the warfare that was an important part of the River and Delta Yuman cultures (see pl. 45). Mastamho's instructions on the use of the war club gave rise to a type of formal combat that was unknown among other Indian groups in the Southwest. While Apaches, Navajos, O'odham, Yavapais, and others conducted raids in order to take booty or exact revenge for previous wrongs, these activities were not like the ritualized contests among the Mohaves, Quechan, Cocopas, and Maricopas.

The Yumans, represented in this description by the Quechan, recognized two types of warfare: the Axwé Omán, a small surprise raid to take captives or rekindle hostility; and the Axwé Hay'áig, or a major, formal battle. The latter often involved as many as several hundred warriors led by a *kwanami,* a war chief, who had dreamed of a successful battle. A Quechan war party, usually joined by allies from other tribes, would journey several days across the desert to attack a Maricopa village. Unlike surprise raids to take captives, advance notice was often given to the adversaries so that they could gather their forces. In some cases the weapon of choice for the battle was even specified beforehand: the bow and arrow, the mesquite war club, or the spear.

As the battle began, the assembled warriors squared off in two opposing lines. The kwanami of each group stood before the rows of warriors and hurled insults at their foes. Then the two leaders engaged in hand-to-hand combat with war clubs. When one was defeated, a bloody melee began, with scores of warriors fighting to the death, most often by crushing their opponent's skull with a heavy mesquite club. The battles often continued

Cocopa men and an Apache woman who was taken captive as a child. Raised as a Cocopa, she married a Cocopa man. (Photograph by DeLancey Gill, Smithsonian Institution, National Anthropological Archives)

for several hours until no one was left standing on one side.

The last such battle occurred in 1857 at a Maricopa village southeast of Tempe. A group of about 300 warriors included a core of Quechan joined by Mohaves, Yavapais, and perhaps a handful of Tonto Apaches. At dawn they surprised the people living in a small cluster of Maricopa homes and killed a few residents. But the Maricopas and their neighboring allies, the Pimas (Akimel O'odham), were ready, quickly gathering as many as a thousand men to protect their homes. At that point the Yavapais, Apaches, and many of the Mohaves fled, leaving about a hundred Quechan and Mohaves to fight to the death. Within an hour, no more than four or five were left alive.

From the 1830s until 1857, these pitched battles occurred regularly, if not frequently. While the carnage was not always as devastating as the 1857 battle, the loss of life was significant. The constant threat of warfare also caused an uneasiness to pervade the relations between the people of the Colorado River. Accounts of earlier Spanish explorers routinely note that guides and interpreters refused to enter the territories of groups outside their own alliance despite the protection of well-armed Spanish soldiers. Families were always on the lookout for surprise raids, and there are reports that the Mohaves had a system of sentries who kept constant watch for Maricopa or Cocopa raiding parties.

Intertribal Warfare and Alliances

The Quechan and Maricopas were at the center of the two opposing military alliances, which involved more than fifteen groups in southern Arizona and California. The Quechan league included the Mohaves, Yavapais, Kamias, Chemehuevis, Hiacid O'odham, and western Tohono O'odham. Peripheral friendships included the Western Apaches and other California Indians. The Maricopas counted among their allies the Cocopas, Halchidhomas, Hualapais, Havasupais, Kavelchadom, Pimas (Akimel O'odham), and eastern Tohono O'odham. These alliances were so important to the region—affecting politics, social relations, economics, and warfare—that ethnohistorians Henry Dobyns, Paul Ezell, and Greta Ezell have compared them to the NATO and Warsaw Pact alliances that dominated Cold War politics in the latter half of the twentieth century (Dobyns, Ezell, and Ezell 1963).

Although information for the period prior to the 1850s is sketchy, it appears that these alliances were fairly stable and persisted over generations. As early as 1774, Garcés noted that the western Tohono O'odham were "hostile to the Halyikwamai, but are ancient friends of the Quechan, whom they have aided in their campaigns" (Forbes 1965:81). But Garcés and other Spaniards, like Lieutenant Michler eight decades later, were not passive observers in the relations between the peoples of the Colorado River. Their actions affected the balance of power in the region and perhaps exerted pressures that increased warfare between the tribes.

As Spanish priests and soldiers made more frequent forays into southern Arizona in the eighteenth century, they developed better relations with some tribes than with others. For example, the Halchidhomas seemed quite willing on several occasions to serve as guides and interpreters for Spaniards along the Halchidhoma Trail, which led from the Gila River overland to the Colorado. The Jesuit missionary Eusebio Francisco Kino reported in 1700 that the chief of the Halchidhomas had sent him a string of twenty-five blue shells as a gift, perhaps to curry favor with the Spaniards, whose military capabilities were well known. Since the Halchidhomas had recently been forced by the Quechan to leave their lands to the south, it is possible that they were seeking to strengthen their weakened position by forming new alliances.

By 1776 the Spaniards were actively seeking to manipulate relations in the region. In an attempt to ensure safe passage along the Colorado as part of a land route to California, they arranged a truce between the Quechan and Halchidhomas. The truce apparently held, with some violations, until 1781, when the Quechan killed the Spanish priests, soldiers, and settlers living among them. Perhaps viewing the Halchidhomas as allies of the Spaniards, the Quechan renewed warfare against them with a vengeance. Through death and the taking of captives for sale as slaves, the population of the Halchidhomas was reduced to a mere handful of families. By the 1830s the few remaining Halchidhomas had left the river for good.

Father Pedro Font commented in 1775 that "This commerce in

Nixoras [slaves], so unjust, is the reason why they have been so bloody in their wars" (Forbes 1965:77). By this time, slave raiding had become an important motivation for the warfare described above as Axwé Omán. Some ethnohistorians speculate that slave raiding was unknown prior to the Spanish era. When Spaniards arrived with impressive new goods to trade, such as metal knives and tailored clothing, Indians soon found that there was great demand for captive women and children, who were taken to the south and sold into a life of servitude. Slave raiding on local enemies represented a convenient source of wealth that fit readily into cultural patterns that placed warfare in high esteem. With the growing presence of Mexican soldiers in the region and increased demand by Mexicans for slaves, the pressure cooker of intertribal relations heated up considerably in the 1820s and 1830s.

The Arrival of the Anglo Americans

In the late 1840s, the small trickle of outsiders passing through the Colorado River region swelled to a flood, with devastating results for the Yuman peoples. During the war with Mexico of 1846–48, u.s. Army troops began to cross westward into California. Despite the fact that Mexico had never occupied or controlled the Colorado and Gila River basins, these lands were ceded to the United States through the 1848 Treaty of Guadalupe Hidalgo and the Gadsden Purchase of 1854. Even before the war, American trappers and prospectors had made frequent visits to the area. When the news of the discovery of gold in California spread in 1849, however, hordes of immigrants from both the United States and Mexico began the trek to the goldfields.

Along the Colorado River, the area of greatest impact for the tens of thousands of overland travelers was the Yuma Crossing at the end of the Gila Trail—the heart of Quechan territory. Like a thousand-mile-wide funnel, westward trails from Texas to Missouri led to this single ford across the river. The forty-niners usually started their journeys poorly prepared and arrived at the river hungry, bedraggled, often hostile, and heavily armed. In addition, they were accompanied by hundreds of famished horses and mules, which made short work of the mesquite beans that were an important food source for the Quechan.

Forty-niners' accounts indicate that the Quechan attempted to be generous hosts to these uninvited visitors. They bartered valuable produce from their depleted stores, often for items such as a cast-off shirt that was of little real value to its owner. The Quechan also performed the dangerous job of ferrying passengers, freight, and livestock across the river, a trade they had monopolized since the Spaniards first entered the territory (see pl. 46). When a horse drowned (providing a convenient meal for a hungry Quechan family), the temper of the owner flared, ignoring the fact that the same horse may have eaten enough mesquite beans to feed that family for a week. Soon it became routine procedure for Quechan ferrying horses to swim the entire distance in the sights of the owner's rifle to insure against accidents.

By the end of 1849, the situation had degenerated. The travelers generally had no regard for the property or person of their hosts. Several armed confrontations broke out, and murders and theft were perpetrated by both sides. A new source of conflict arose when Mexican and American entrepreneurs built makeshift ferries to compete with the Quechan. A small contingent of army troops arrived in 1849, and this served to reduce tensions somewhat. However, by November of that year Lieutenant Cave J. Couts was operating his own ferry across the river.

After two unsuccessful crops in 1850 and 1851, the attention of the Quechan was almost totally focused on feeding themselves and dealing with the increasing presence of the Americans. They joined with their traditional enemies the Cocopas and members of several tribes from southern California in an ill-fated uprising in 1851. The army was forced to abandon its camp at Yuma, but it later crushed the insurgents in California. In 1852 a new detachment of four hundred troops arrived at Camp Yuma. Major Samuel Heintzelman led a three-month campaign directed primarily at burning houses and destroying fields. After a showdown with a combined force of Quechan, Mohaves, and Yavapais, the army prevailed and a truce was arranged. The Quechan were forced to submit to these intruders on their lands.

With a lesser military presence on the northern reaches of the river, the Mohaves were able to hold out a bit longer. Following the incidents in 1852, Mohave combatants apparently returned home and resumed their normal activities. For them, contacts with white immigrants and soldiers were infrequent during the early 1850s. In 1858, however, the Mohaves attacked the first large party of settlers who attempted to cross the river near Hardyville. In urging the attack, one of the kwanami warned: "If we let the whites come and live here, they will take your wives. They will take your children and carry them away and sell them. They will do that until there are no Mohave here" (Kroeber and Kroeber 1973:12).

In 1859, hundreds of troops were ordered into the Mohave Valley to take hostages and destroy fields and homes. The campaign included several battles that pitted war clubs against rifles. After the loss of dozens of warriors, the Mohaves finally admitted defeat. At the conclusion of hostilities, Major Lewis Armistead admonished them: "If you do [fight], we will take away your land and give it to white people" (Kroeber and Kroeber 1973:33).

Reservation Lands

Despite the Mohaves' agreement to lay down their arms, Major Armistead's warning still came true. The federal government formally took possession of the bulk of Yuman lands in 1865 by creating a relatively small reservation in the Colorado River Valley north of the mining community of La Paz between modern Blythe and Parker. In arguing for these dedicated lands, the superintendent of Indian affairs for Arizona, Charles Poston, stated, "Their resources from game, fish, and wild fruits have been very much curtailed by the influx of Americans, and it would be dangerous for them to visit their former hunting grounds" (Fontana 1958:5–6).

Congress set aside a mere 75,000 acres to create the Colorado River Indian Reservation for the use of the Delta, River, and Upland Yumans, who had previously used lands that amounted to about one-third of the state of Arizona. In addition, the reservation was to be used by "the Indians of [the Colorado] river and

The war leader Xantapacheria (seated, right center) with an all-Maricopa volunteer unit during the Civil War. The Arizona Volunteers, composed largely of Maricopas, Akimel O'odham, and Mexicans from Sonora, were far more effective fighters against the Apaches and Yavapais than were U.S. troops, but they were disbanded after the war ended. (Barry Goldwater Collection, Arizona Historical Foundation)

its tributaries" (Fontana 1963:167), a group that could include all of the Indians of Arizona and parts of Utah, Colorado, New Mexico, and California!

By the 1870s, the lives of the Colorado River peoples had been totally disrupted. Relations between the different groups continued, but interactions with whites and federal agencies dominated their lives. The government attempted to settle Quechan, Hualapais, and Yavapais on the Colorado River Indian Reservation, but these efforts were unsuccessful. It was unreasonable, not to mention unjust, to expect these peoples to leave their homelands and settle on the traditional lands of the Mohaves in such close quarters. By 1890 only the Mohaves and Chemehuevis remained on the reservation. The Fort Mohave Reservation was established in 1870, and eventually other reservations were set aside for the Cocopas, Quechan, and Maricopas.

The Resettlement of Hopis and Navajos

Although 75,000 acres was not a large parcel in comparison to the original native territories, by 1900 the white citizens of Arizona had begun to view these prime agricultural lands as much too large a tract to be wasted on a handful of Indians. The Dawes Severalty Act of 1887 paved the way for outside takeovers by allowing small plots of reservation lands to be deeded to individuals. In some cases the "surplus" land was sold to outsiders at auction. In 1908, Congress auctioned a portion of the Colorado River Indian Reservation and established the town of Parker within its boundaries.

This move did not satisfy the land speculators and farmers, who kept a watchful eye on the political situation. In 1915 the Arizona legislature petitioned Congress to open the Colorado River Indian Reservation to land sales, stating that "the surplus Indian lands . . . are highly desirable as prospective farms, and hundreds of energetic and enterprising citizens of this State alone are awaiting the opportunity to secure tracts of lands for the purpose of making their homes thereon" (Fontana 1963: 169).

In 1931 the BIA reservation superintendent, C. H. Gensler, backed the sale of reservation lands as a means of paying for needed irrigation projects. Shortly thereafter, however, he proposed the colonization of the reservation by Indians from other parts of the state, citing the "tributaries" clause of the original 1865 legislation. The idea was received warmly by the commissioner of Indian affairs, and in 1935 a group of Navajos began inspecting possible homesites along the river.

The members of the Colorado River Indian Reservation had mixed feelings about these developments. On the one hand, they were understandably reluctant to turn over a part of their homeland to new residents of different cultures who would have a large voice in running the affairs of the tribal council. On the other hand, they responded with compassion to appeals on behalf of Indians in Arizona who were feeling the effects of the nationwide depression. Indian colonization would also help ward off an auction of valuable lands to white farmers. The

completion of Hoover Dam in 1936 promised the future development of extensive tracts of irrigated farmlands, and pressures on reservation lands were expected to increase. BIA officials
continually referred to this threat in negotiations with the tribal
council.

The Colorado River Indian Tribe passed Ordinance Number 5
in 1945, dedicating the larger, southern portion of the reservation to new colonists and providing perpetual land assignments
and full tribal membership to immigrants. Within six years, 148
Hopi and Navajo families had arrived, nearly equaling the existing population of approximately 1,200 Mohaves and Chemehuevis.

Feeling very nervous about the whole series of events, the
tribal council rescinded the ordinance in 1952 while they still
had a chance. The rapid influx of settlers promised to make them
a political minority in a very short time. The full impact of the
loss of the southern portion of the reservation became known as
some new settlers wound up with larger landholdings than tribal
members who had been on the reservation since birth. In some
cases this situation resulted in strong resentment, especially
among Mohaves who had not received allotted lands. Time has
tempered most of these hard feelings.

Cooperative Interaction for the Future

Today the Colorado River Indian Reservation is a thriving multi-
ethnic community consisting primarily of people of Mohave,
Chemehuevi, Hopi and Navajo descent. Through marriage,
members of other Arizona Indian nations and even Anglo Americans and Hispanics live on the reservation.

The Colorado River Indian Tribes Museum and Library preserves the heritage of all four major Indian culture groups. It has
an active program devoted to collecting and preserving artifacts
and documents, holding language and craft classes, conducting
oral history interviews, recording archaeological sites, and spon-
soring a tribal elders' group. Composed of individuals who are
especially knowledgeable about their cultural traditions and
spiritual matters, the group meets regularly. They share their
knowledge and experiences, discuss current matters such as

A home on the Quechan reservation in about 1900. (California Historical Society, Ticor Title Insurance Collection, University of Southern California Library)

boundary issues and legal cases, and record cultural information such as place names, archaeological sites, and their language. Since 1984 the tribe has videotaped the elders' meetings, producing a lasting record that will speak to future generations with a compelling voice about important cultural values.

Relations with non-Indian peoples living along the Colorado River have also been productive and usually cordial. For decades the Colorado River Indian Tribe has leased a portion of its fertile tribal lands to outside farmers. It has also enjoyed a cooperative relationship with the town of Parker, a commercial center that serves the reservation as well as the vacationing boaters that flock to the river for waterskiing and fishing. Many local merchants rely heavily on the business generated by their Mohave, Chemehuevi, Hopi, and Navajo neighbors.

Several years ago, however, these relations were strained by an issue that had its roots deep in the history of the region. The dispute arose in the early 1980s when the tribe planned to develop certain lots that it owned in Parker. City officials asserted that all projects must comply with city building codes, while the tribe maintained that since the lots were tribal lands, only tribal

THE COCOPA GAME OF PEÓN

Nancy J. Parezo

Like many other Native American groups, the Cocopas of the lower Colorado River are actively searching for ways to preserve their cultural traditions. One such effort is an oral history project in which members of the Cocopa tribe's cultural preservation task force provide tape recorders to elders so that they can relate stories and incidents in their lives whenever they have the chance. These stories are then transcribed and sometimes translated and placed in the new cultural archive and museum.

Another endeavor is a revival of the gambling game of peón (*micuul*), a term of Spanish origin. Two teams composed of four players each seat themselves on the ground across from one another in a north-south direction. A group of singers stands behind the players while the game progresses and sings appropriate songs. The members of each team hold one arrow-weed stick painted black and one white egret bone. Each team also has a blanket that the members hold in their mouths while the sticks are being arranged. The object of the game is to guess which hand holds the white bone.

Both spectators and players bet beds, beads, hair belts, money, or baskets on the outcome of the game. The game used to be played only by men, but today both men and women participate. The referee, who has a helper to lead the singing, holds fifteen counters. The guesser from one team points between any two players on the other team to indicate that those seated on the north side of where he or she points hold the white bones in their northside hands while those to the south hold the bones in their southside hands.

If he or she guesses correctly, the guessing team wins a counter. If the guess is incorrect, the other team receives a counter. When the fifteen counters are gone, the referee pays out the winner from the losing team's supply. When one side wins all the sticks and counters, it wins the jackpot. The winning team and the community members who supported it split the winnings. According to tribal council member Will Ortega, who spoke to KUAT radio reporter Bobbi Clark on the program *Many Voices: Today's Native Americans* in 1994, "It [peón] is one of our popular games, and some of the elders are using it to teach our people the traditional ways of things. . . . A lot of people are getting interested in it, and the language that comes in there, too."

laws and building codes applied. The situation heated up when Parker officials cut off electricity and water service to tribal buildings in town. As then tribal vice-chairman Elliott Booth said, "The community of Parker has an attitude problem" (Helliker 1988:36). The tribe responded by filing a federal lawsuit in 1983, claiming the same jurisdiction over these lands that it holds over reservation lands.

The question of tribal jurisdiction in Parker goes back to the formation of the town. When the Southern Pacific Railroad built a new line to Los Angeles in 1908, Congress removed 900 acres from the Colorado River Indian Reservation for a town site at the new river crossing. Building lots were auctioned, with the proceeds and the unsold lots turned back to the tribe. Thus the tribe owned lands within the town of Parker, but Parker was within the original Colorado River Indian Reservation.

As the lawsuit wended it way through u.s. district court, fears and suspicions escalated as relations worsened. The tribe offered to meet to work out a compromise, but Parker officials refused to negotiate unless their jurisdictional authority was recognized prior to any negotiations. "That's insulting," said William Lovell, an attorney for the tribe. "They want us to totally capitulate and call it a compromise" (Helliker 1988:38). The tribe then called for a boycott of businesses owned by the Parker mayor and police chief. Meanwhile, many Parker residents feared that if the tribe won the case, they would claim jurisdiction and even ownership of all town lands that had originally been part of the reservation.

On January 17, 1989, Judge Roger Strand handed down a ruling in the case, establishing that "the tribally owned lots of the Town of Parker now are, and always have been part of the Colorado River Indian Reservation" and that the town of Parker "does not have the authority to impose or enforce any building or zoning ordinances with respect to this tribal property" (*Colorado River Indian Tribes* v. *Town of Parker*). Although not a clear-cut victory for the tribe, the ruling also did not have the far-reaching effects that some may have feared or hoped.

While the lawsuit is still a hot issue in Parker, relations are returning to normal between the Colorado River Indian Tribes and their neighbors. Despite ethnic and cultural differences, the Yuman-speaking Mohaves, Quechan, Cocopas, and Maricopas have a long history of coexisting with their neighbors along the Colorado River. The different ethnic groups of the Colorado River Indian Tribe hash out their mutual problems in council meetings, enter into joint agricultural ventures, occasionally trade a Hopi kachina for Mohave beaded jewelry, and marry spouses from other groups. Although the players and the dynamics may have changed, the different peoples of the Colorado River region maintain their rich tradition of social interaction.

Plate 39. Seri seaweed dolls. The sargasso seaweed doll at left was made by Sara Villalobos in 1941. The eelgrass doll at right was made by Ramona Casanova around 1972. (Photograph by Helga Teiwes, Arizona State Museum)

Plate 38. Pre-Seri or Seri "eggshell" pottery sherds and figurines. The sherds date between A.D. 700 and 1800, and the figurines to before 1900. The lack of well-dated archaeological sites in Seri territory has hindered the assignment of more specific dates to such objects. (Photograph by Helga Teiwes, Arizona State Museum)

Plate 40. Seri carrying pouches. Clockwise from top, they are a bag made of a pelican throat, before 1976; a canteen made of a sea turtle bladder, before 1941; and a bag made from a turtle stomach, before 1977. (Photograph by Helga Teiwes, Arizona State Museum)

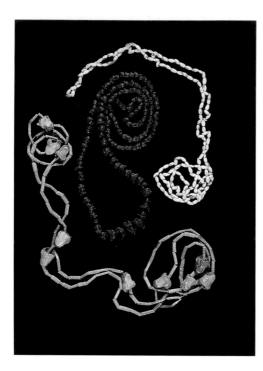

Plate 41. Seri necklaces. From top to bottom, they are necklaces of Olivella shells, made in 1940; octopus suction cups, made before 1960; and lobster antennae and eyes, made before 1960. (Photograph by Helga Teiwes, Arizona State Museum)

Plate 42. Seri coiled basketry. They are, from left, bowls made before 1940 and before 1950, and a jar made by Elvira Valenzuela Félix before 1961. (Photograph by Helga Teiwes, Arizona State Museum)

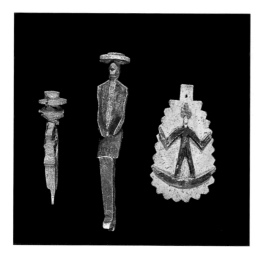

Plate 43. Seri santos. They are, left to right, a Janus male figure, a male figure, and a male figure on a crescent, all made before 1941. (Photograph by Helga Teiwes, Arizona State Museum)

Plate 44. Mohave figures, made around 1930. In the late nineteenth century, Quechan and Mohave women began making ceramic dolls with horsetail or human hair and dressed in traditional clothing, beadwork, and body painting styles. These were first sold to rail travelers passing through the Colorado River towns of Yuma, Arizona, and Needles, California.

Plate 45. Mohave warriors in the late 1850s. (Illustration by Heinrich Balduin Möllhausen, Museum für Völkerkunde, Berlin)

Plate 46. A Quechan ferry on the Colorado River in about 1854. (Illustration by J. J. Young)

Plate 47. A Colorado River Indian Tribe grouping. Clockwise from top right, they are a Hopi flower kachina, made by Hopi Ron Yava in 1988; a Mohave ceramic jar, made by Elmer Gates before 1976; a Chemehuevi basketry jar, made before 1940; and a Navajo rug, made before 1989 on the Colorado River Indian Tribe reservation.

(Photograph by Helga Teiwes, Arizona State Museum)

Plate 48. Colorado River Yuman tourist crafts. They are, from top, a Mohave toad effigy, made by Annie McCord Fields in 1961; a Quechan flute, made before 1903; and a Mohave necklace, made before 1937. (Photograph by Helga Teiwes, Arizona State Museum)

Plate 49. Examples of Maricopa pottery. They are, left to right, a jar and two effigy jars, all made before 1907. (Photograph by Helga Teiwes, Arizona State Museum)

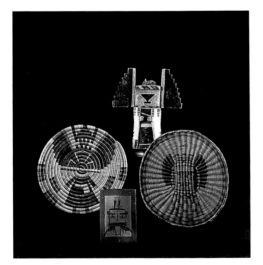

Plate 50. Hopi art using Kachina motifs, which express central concepts of Hopivotskwani. At left is a coiled plaque with an Angwusnasomtaqa (Crow Mother) motif, made before 1940; at top center is a Crow Mother kachina, carved around 1900; at bottom center is a pottery tile depicting Kuwanhehey'a Kachina, signed by "te-wang-i-nema," purchased at Polacca on First Mesa before 1941. At right is a wicker plaque with a Crow Mother motif, made before 1916 at Oraibi. (Photograph by Helga Teiwes, Arizona State Museum)

Plate 51. A Hopi belt from Walpi, on First Mesa, in 1919 and a brocaded sash from Oraibi, on Third Mesa, from before 1915. Hopi men produce woven textiles, many of which have designs associated with Kachina spirits, fertility, and rain. (Photograph by Helga Teiwes, Arizona State Museum)

Plate 52. A Hopi *tableta,* made about 1950. Women wear tabletas in social dances such as the Butterfly Dance. They are also worn by some Kachina dancers. (Photograph by Helga Teiwes, Arizona State Museum)

Plate 53. A Hano polychrome jar. The most famous of the many Hano artists was Nampeyo. This jar was made by members of the Nampeyo family, possibly a collaboration between Nampeyo and her daughter Fannie in the early 1900s. (Photograph by Helga Teiwes, Arizona State Museum)

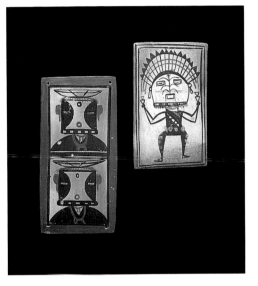

Plate 54. Hopi tiles. Trader Thomas Keam suggested that Hopi potters produce tiles for Anglo American fireplaces. These were very successful and became prominent in Hopi artists' repertoires. At left is a double tile signed by Te-Wang-Nema and depicting *kipokkatsinam* (raider katsinam), made before 1941. At right is the Hopi version of the Navajo Talking God, the Tasaf Kachina, painted before 1919 on First Mesa. (Photograph by Helga Teiwes, Arizona State Museum)

Plate 55. A contemporary Hopi jar from 1971. Elizabeth White (Polingaysi) was a distinguished Hopi artist from Third Mesa. Her innovative yellowware with sculptured corn emphasizes the importance of corn as a central symbol of Hopivotskwani. (Photograph by Helga Teiwes, Arizona State Museum)

Plate 56. A Niman kachina carved by Otto Pentewa of New Oraibi before 1959. Kachina dolls have been a part of Hopi culture for centuries. They are given as gifts to Hopi female children. (Photograph by Helga Teiwes, Arizona State Museum)

Plate 57. Clark Tenakhongva carving a *tihu* (kachina doll) in 1994. Hopis walk many paths. A cattle rancher, tihu carver, and educator, Clark Tenakhongva also serves his community as president of the Polacca Day School Board and as a member of the Hopi Board of Education and the Hopi High School Board. His research into older carving styles has helped renew interest in traditional forms of kachina dolls. (Photograph by Russell Varineau, Arizona State Museum)

THE HOPIS

HOPIVOTSKWANI, THE HOPI PATH OF LIFE

Nancy J. Parezo

According to Hopi belief, all life is a journey and each group of people have a path. For Hopis, it is part of Hopivotskwani, the Hopi Path of Life. As people emerged into this world, the Hopi-tuh Shi-nu-mu, the Peaceful People, made a choice to follow the Hopi Path. According to anthropologist and Hopi judge Emory Sekaquaptewa, Hopivotskwani includes every part of Hopi society and culture: kinship and family, religious ceremonies and beliefs, political organization, concepts about ownership, farming, world view, art, and even ideas about the prehistoric ruins of the Southwest. It is a comprehensive philosophy that touches the lives of most Hopis on a daily basis. Hopivotskwani gives Hopis a strong sense of security in an often-troubled world by relating the past to the present and insuring cooperation with supernaturals who provide rain, abundant crops, health, and well-being. Hopis do not expect outsiders to follow their laws and traditions, and in turn they feel they should not be forced to follow others' paths. Hopi history is in part an evolving trail punctuated by episodes in which the Hopis have been tested by hardship and outside threats to their way of life, including interaction with peoples who have tried to force them to follow other trails. Their response to these situations has been to isolate themselves and reassert Hopivotskwani.

A basic tenet of Hopi philosophy is that humans live in harmony with nature. In this view, religion permeates all aspects of an inextricably interwoven life. Nature, gods, spirits, animals, plants, the land, and people are one in an unchangeable relationship. This philosophy and relationship is community based.

Each Hopi village is ritually considered to be the center of the world, and each respects the others' independence. Bounded on four quarters by holy mountains, this world contains within it places where contact may be made with spirits through sacred rites performed to ensure prosperity and the proper functioning of Hopivotskwani. For example, throughout the growing season, prayers for germination, plant growth, and maturation are offered and thanks given for prayers "answered and unanswered." Lean and bountiful years are attributed to the faulty or proper observance of ritual, a prerequisite for which is that every individual maintain a good heart.

Hopivotskwani is also reflected in attitudes toward the land. The land on which Hopis live and over which their ancestors have walked is sacred, having been designated as Hopi Land by supernaturals. This land cannot be bought or sold, and Hopis cannot give it away, for they are entrusted with preserving it. Like the Yaquis and the Pais, the Hopis have developed a body of tradition that underlies and reinforces this belief. As Hopi scholar Hartman Lomawaima has remarked, "At Hopi, the oral tradition is still the single most viable means of transmitting information to new generations. Religious and secular knowledge is held by individuals and collectives (clans and religious societies) and is transmitted according to a specific plan" (1989: 94–95). It is in this way that Hopivotskwani remains vital and Hopi individuals learn about the essential parts of their history.

The Early Path: Footprints of the Clans

Hopi history begins with the emergence of the ancestors into the land of Masauwu. The ancestors then set out to discover the world, to find the place where they could live productively and properly, free from drought and imperfection, in spiritual harmony with nature. But as mortals they could not all agree on the same path, and they split into many groups, or clans, each descended from a respected woman, each perfecting its own version of Hopivotskwani as they traveled. Paahana, Elder Brother, told the people that he would know the path each group took by following their footprints. These "footprints" are the prehistoric ruins of the northern Southwest. According to Hopi oral tradi-

A Hopi farmer in about 1900. As part of the agreement to follow the way of life of Masauwu, farming is a spiritual rehearsal and symbolic representation of one's own life. To grow successful crops, a man must be humble, hardworking, and willing to endure hardship. (Photographer unknown, Arizona State Museum)

tions, these sites are the places where clans lived before arriving at the Hopi Mesas. The prehistoric communities at Mesa Verde, for example, are regarded as the former home of the Badger clan, while Wupatki is the home of the Water clan. According to Albert Yava, the footprints of the Snake and Horn clans can be found to the north at Tokonave (Navajo Mountain) among the

EMERGENCE TO THE FOURTH WORLD

Nancy J. Parezo

Hopi origin stories relate the principles of Hopivotskwani and provide a means for Hopis to link the present with the past. But Hopi is not a homogeneous society. There are many different versions of the origin legends, just as there are versions of oral traditions, ceremonies and ritual equipment. The following is a brief summary of the Hopis' emergence to the present world adapted from Ekkert Malotki and Michael Lomatuway'ma's book *Stories of Maasaw, a Hopi God* (1987) and Harold S. Courlander's *The Fourth World of the Hopis* (1971). The emergence set the pattern for Hopi migrations and quests that are discernible throughout Hopi history.

In the beginning, Tawa, the sun spirit and father, impregnated Mother Earth, who gave birth to living things. These people, animals, and insects lived in the underworld, where they tried to live the Hopi Way but were not able to understand the meaning of life and became mired in corruption and strife caused by sorcerers. Upon hearing footsteps above and the words of Hummingbird and Spider Grandmother, a brave group decided to leave the *koyaanisqatsi,* or "crazy life," behind. They sent out birds to find a way to ascend and explore the unknown land. Catbird finally succeeded in reaching the *sipapuni,* a hole in the sky, and found Masauwu.

Masauwu, the Spirit of the Dead, gave the people fire and permitted them to settle on his land. The Fire People, who have a special relationship with Masauwu, emerged first, because they agreed to assume the responsibility of leading the others to their final destination. Chipmunk helped the good people climb to the si-

papuni through a tall reed. As the people were resting before their journey, Spider Grandmother said, "The journey will be long and difficult. When we reach the Upper World, that will be only a beginning. Things there are not like things here. You will discover new ways of doing things. During the journey you must try to discover the meaning of life and learn to distinguish good from evil. Tawa did not intend for you to live in the midst of chaos and dissension."

On the fourth day, the good people ascended, following the boy warrior gods Pokanghoya and Polongahoya. Yavpa, the Mockingbird, addressed each person as he or she set foot in the Upper World, saying, "You shall be a Supai and speak the Supai language," or "You shall be a Paiute," or "You shall be a Hopi." In this way, tribes were established and made distinct.

Then the boy warrior gods made the land from the soft mud, grass, and trees. But the Upper World was dark, so Spider Grandmother helped the people make the sun and moon through their power of singing. When the world was bright and warm, the people chose their corn. Masauwu and Yavpa arranged six ears of colored corn in a circle and asked the people to chose the ear they wanted. Mockingbird said, "Each of these ears brings with it a way of life and language. The one who chooses the yellow ear will have a life full of enjoyment and prosperity, but his span of life will be small. The short blue ear will bring a life full of work and hardship, but the years will be many." The Navajos took the yellow, the Supais the speckled, the Comanches the red, the Utes the flint corn, the

Apaches the *kwakwi* grass, and the Hopis, the last to choose, took the stubby blue corn *sowi-wa*. The Hopis knew that this ear would last forever, and Masauwu and Mockingbird congratulated them. The Hopis said, "We will have a life of hardship, but it will be a long-lasting life. Other tribes will perish, but we, the Hopis, will survive all adversities."

Thus the Hopis chose Hopivotskwani. With their decision they displayed humbleness, and Masauwu and Mockingbird instructed them on how to live. They incised rules on two stone tablets that were to be used to help the Hopis settle their disputes. Masauwu and Mockingbird gave one tablet to each of two brothers chosen to be leaders. Younger Brother was to be Hopi, and Older Brother, Paahana, the whites. Then Masauwu told everyone to leave sipapuni and migrate across the continent, each group in a different direction, in order to learn their way of life before reaching their final destination. So the Hopis went in search of their promised land.

Utes and Paiutes; the footprints of the Bear, Bluebird, Spruce, and Gopher clans lie to the west; those of the Coyote clan, east to the Plains; and those of the Sand, Sun, Rabbit, and Tobacco clans, to the south and west at Palatkwa (Yava 1978:36).

Archaeological evidence supports the Hopi claims of connection to many known sites. Archaeologists believe the Hopis have direct ties to the "Kayenta Anasazi" (a term Hopis consider disrespectful) or the Western Pueblos, the prehistoric peoples who lived in the many "cliff dwellings," hamlets, and villages of northeastern Arizona. (The Hopis call these ancestors Hisatsinom.) The Homol'ovi pueblos, seven large ruins about sixty miles south of Hopi, were major cotton-producing villages with strong trade and religious ties to Hopi. Settled by people from eastern Arizona, the Homol'ovi area was abandoned in the early 1400s, and its inhabitants probably joined existing Hopi villages. Scientific evidence for this interaction and migration is strong: neutron activation analyses of ceramics found at Homol'ovi II (the largest pueblo, consisting of more than 1,000 rooms) indicate that yellow pottery was made at Hopi villages and traded to the Homol'ovi region in return for cotton (Bishop et al. 1988). Finally, the representation of Kachinas on rock art, murals, and pottery at Homol'ovi in the late 1300s gives evidence of religious ties to the "Kachina Cult," which was so highly developed among the Hopis (Adams 1991). Trade was also extensive. The presence of the Hopis' yellow-ware pottery as far south as the Tucson Basin attests to their extensive interaction networks.

The population at the Hopi villages grew during the thirteenth to fifteenth centuries. As many Hisatsinom sites were abandoned, clans moved to their prophesied home on the mesas. These kinship groups were drawn to the area because it was Sichdukwi (Flower Mound), a promised land where they could follow Hopivotskwani. On Third Mesa it is said that the Bear clan was the first to arrive. Its members founded Oraibi by A.D. 1150. According to Emory Sekaquaptewa, Hopi legend states that Oraibi was founded after a conflict between two brothers over a woman: "Matcita and Yahoya are said to have stood on the mesa midway between Shungopavi and Oraibi, and entered into a covenant which stipulated that the one who increased 'his children' (society) over the other should prevail. The prevailing chief would claim all the land between them, the boundary running through the middle of the loser's village" (Sekaquaptewa 1972:239).

When other clans came, they had to prove their knowledge of Hopivotskwani to the Bear clan, demonstrate the value of their rituals, and agree to share them before they were allowed to settle and receive land. For example, the Badger clan brought the Powamuya ceremony, which has the power to call on Kachina spirits to assist in the germination of crops.

This ability to work with the Kachina spirits is an important basis of Hopi society (see pl. 50). The Kachinas reside in a spirit world, which mirrors the world we know. They represent the spiritual essence of people, animals, plants, and objects, and they serve as spiritual guides and messengers. When people die, they become Kachinas and are seen in clouds. Complex and many-faceted spirits, Kachinas can intercede with deities on behalf of the Hopis. From the winter solstice until July, they visit the Hopi villages and participate in rituals. Caring deeply about the people, they dance for them, bring gifts, and teach the Hopis arts and industries as well as proper behavior. At other times they dwell in mountains, springs, and lakes. As benevolent beings, Kachinas possess the power to sustain life by bringing rain and ensuring fertility and well-being for people, animals, and crops. As long as the Hopis follow Hopivotskwani and "their hearts are right," the Kachinas will come and bless the land and people.

Ceremonial control of important knowledge, behavior, and ritual early formed the important cornerstone of Hopi society that continues to this day. A set of religious societies (concerned with rain, health and curing, ritual clowning, social control, and war) developed prehistorically and continued to evolve in the historical period. These societies provided a body of ceremonial and social checks and balances. Each Hopi ceremony is controlled by a specific clan, usually the one that brought the ritual to the Hopis (a relationship that is symbolized by a special ear of corn; see pl. 55). The male and female clan leaders control the power of the ceremony, are responsible for its success, and maintain its sacred altar and objects. Rituals, however, are performed by members of a community-wide religious society. While clan membership is determined at birth, anyone can join a religious

THE HOMOL'OVI RUINS OF NORTHEASTERN ARIZONA

E. Charles Adams

The word *Homol'ovi* is Hopi and means "place of the mounds or rolling hills." It is used to describe an area sixty miles south of the present Hopi villages in northeastern Arizona. This area lies along the Little Colorado River near present-day Winslow and was used historically by the Hopis for crossing the river and as a place to catch turtles for use in some of their ceremonies. The Homol'ovi area is also of interest to the Hopis and archaeologists alike for its ancient ruins, which all agree are ancestral to the villages of the modern people. The Hopis identify several of their clans, groups of related people who traveled or migrated together, that lived at Homol'ovi for a time before arriving at the Hopi villages.

The Homol'ovi ruins consist of seven pueblos: Homol'ovi I, II, III, and IV, Cottonwood Creek Ruin, Chevelon Ruin, and Jackrabbit Ruin. These pueblos are located along a twenty-mile stretch of the river. All were occupied between about A.D. 1250 and 1425. The Homol'ovi sites were first investigated by the Smithsonian Institution archaeologist J. Walter Fewkes in 1896. Fewkes had learned of the existence and location of the ruins from Hopis whom he had been studying and whose ruins he had been excavating the previous year. Fewkes focused his excavations on Homol'ovi I and the Chevelon Ruins, especially on cemetery areas, from which he took perhaps 500 pottery artifacts plus hundreds of other objects.

Fewkes's work preserved part of the archaeological heritage of this country and part of the cultural heritage of the Hopi Indians. His excavations, however, brought the ruins and their contents to the attention of residents of Winslow, railroad workers and passengers, and many others who systematically plundered the pueblos up to 1979. In 1986, five of the pueblos were incorporated into Homolovi Ruins State Park, with the other two monitored by state park personnel.

Recent research by Arizona State Museum archaeologists has shown that, by the mid-1200s, the builders of the Homol'ovi pueblos took advantage of excellent environmental conditions and large labor pools to enable them to farm the broad, fertile floodplain of the Little Colorado River. In addition to growing their staple foods—several varieties of corn, beans and squash—they grew cotton in quantity. Excavations indicate that numerous looms were established in the kivas to weave this cotton into textiles. Cotton seeds and fragments of textiles are plentiful in the excavations. Cotton production probably enabled the Homol'ovi people to trade for various exotic goods, including shells from the Gulf of California, copper bells and macaws from western and northern Mexico, obsidian from the nearby San Francisco Mountains, and pottery from numerous sources.

During the mid-1300s, the largest pueblo, Homol'ovi II, with more than a thousand rooms, was established and came to dominate the area politically. During the late 1200s and early 1300s the cotton trade may have been focused to the east and southeast, but with the establishment of Homol'ovi II this exchange clearly shifted to the north. At contact in the 1500s, the Spanish noted cotton textiles and extensive cotton fields among the Hopis. This pattern seems clearly to have been in place during the occupation of the four Homol'ovi ruins.

By the early 1400s, the Homol'ovi area had been abandoned. Most of the people who had lived there went to villages on the Hopi Mesas. The causes of the abandonment remain uncertain. There are clear indications that topsoil was being lost in surrounding upland areas where agriculture was being practiced and corn would have been grown. Alternating episodes of flooding and drought in the late 1300s and early 1400s may have made the floodplain too unreliable for the production of food or cotton. Whatever the cause, the legacy of the Homol'ovi pueblos and their people has been preserved by their descendants, the Hopi Indians.

society. Thus, a person will have a key role in a ceremony controlled by his or her clan but will participate in other rituals throughout the year as a member of a religious society. Following Hopivotskwani requires one to play many roles.

Encounters with the Spaniards

When Spanish explorers pushed northward from Mexico during the sixteenth century, they encountered Indians living in compact, many-chambered, flat-roofed structures built around plazas. These single- to multistoried settlements surrounded by fields reminded the explorers of the villages of their homeland, so they called them pueblos and the residents Pueblo Indians. Anthropologists today divide the Pueblo Indians into two groups—the Eastern Pueblos, who live along the Rio Grande and its tributaries, and the Western Pueblos, the Hopis, Zunis, Acomas, and Lagunas. All are similar yet distinct.

The Spanish explorer Pedro de Tovar, a lieutenant of Coronado, and seventeen horsemen came to the province of "Tusayan" in 1540 and visited Kawaikuh on Antelope Mesa. After this brief meeting, the Hopis were defeated and the village was partially destroyed. Tovar briefly visited other Hopi villages and found more than 3,000 Hopis living in seven villages located at the foot of mesas near springs: Awatovi, Sikyatki, Walpi, Kawaikuh, Mishongnovi, Shungopavi, and Oraibi. Each village was autonomous and functioned independently in all areas. There was no pan-village tribal government, although the villages could and did act in concert. Internally, each was strongly united and highly organized through an intricate system of matrilineal and matrilocal clans composed of small kinship groups called lineages. Each clan was related to a *wuya* (an animal, supernatural, or plant spirit), from which it derived its name, and each maintained proper reciprocal relationships with the supernatural by vigorously and conscientiously performing religious rituals for which the clan was responsible.

Within the village, the female-based clan system resulted in the construction of blocks of related households adjacent to the home of the matriarch, usually the eldest woman in the senior lineage. Households composed of extended families were the

The village of Walpi on First Mesa in about 1900. For centuries, Hopis have lived in multistoried pueblo communities. (Photograph by C. C. Pierce, University of Southern California)

basic economic units, and labor was judiciously divided between men and women. According to the Hopis, women stabilized the society. They owned the houses, and a woman brought her husband home to the house of her mother upon marriage, later building a house nearby. Women also owned agricultural fields, food, seed for the next year's planting, springs and cisterns, and small irrigated gardens. As day-to-day activities, women carried water up the steep cliffs, built houses, made pottery and baskets, and cared for the family. Men were responsible for government and religion; did the herding, hunting, and farming away from the village; gathered and hauled fuel; spun cotton; and wove. As a basic part of Hopivotskwani, these male and female complementary work patterns in pursuit of the common good continue today.

The history of Hopi and Spanish interaction well illustrates the limited resources of Spain on the northern frontier and the Hopi assertion of Hopivotskwani. The seventeenth century was

WOMEN'S ROLES
THE HEART OF HOPI SOCIETY

Nancy J. Parezo and Bruce Hilpert

Hopi grandmothers, mothers, and their young children are members of the same clan and usually live in adjacent households. They form the backbone of Hopi society. A woman's roles are many.

Mother and Provider

When a woman gives birth, she ensures the future of her clan. Through hard work she provides for her children. Women are responsible for maintaining the household, processing and cooking food, and tending the kitchen gardens.

Owner of Property

Farmlands, houses and other property remain in the clan and are passed from mother to daughter. Women control resource distribution. A husband works his wife's fields and may build a new house but has no claim to ownership. A man whose wife dies moves back to the house of his mother or sister.

Clan Matriarch

If a woman is the senior female member of a clan, she makes many of its decisions. Her household is regarded as the ancestral one for the clan, and she is referred to as the Clan Mother.

Clan Religious Leader

Although a woman does not participate publicly in Kachina rituals, a matriarch passes on to her sons the essential knowledge of the clan's ceremonies. She is also responsible for safeguarding religious equipment.

Member of a Religious Society

A woman represents her clan as a member of religious societies that perform women's ceremonies each fall.

Creator of Beauty

Women are responsible for making baskets and pottery, which are needed for weddings, ceremonies, gifts, and trade.

Grandmother

Elders teach Hopi ways and help grandchildren out of difficulties. They are very affectionate and hold a position of utmost respect in Hopi society.

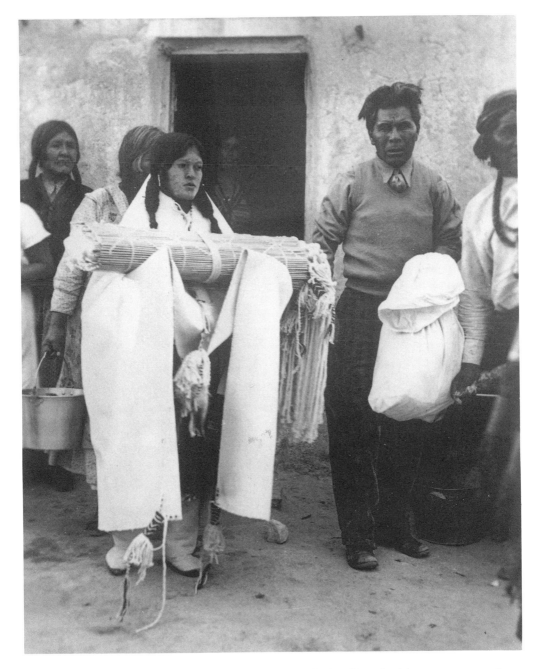

A Hopi bride wearing her wedding garments leaves her new husband's mother's house to return to her mother's home in the 1940s. Hopis are matrilocal; couples live in the home of the bride's mother right after marriage. (Museum of Northern Arizona)

a period of considerable settlement change among all the Puebloan peoples who were becoming Hispanicized. Kawaikuh was
abandoned as a result of Spanish contact, while Sikyatki residents emigrated because the failing water supply in the Jeddito
Valley made agriculture difficult. In general, contact with explorers was brief, intermittent, and variable. When Antonio de
Espejo came in 1583, he spent several days at Awatovi, where he
was not welcomed, but he met a friendly reception elsewhere,
and Hopis guided him in his search for mineral deposits. Juan de
Oñate followed in 1598, trying to consolidate Spanish political
control. While the Hopis submitted without resistance, it was
only a formality, and in their isolated location, the Hopis remained independent, paid no tribute, and were not subject to
Spanish military control.

The Spanish did, however, establish a limited presence beginning in August 1629, when the Franciscans established missions
at Oraibi, Shungopavi, and Awatovi and *visitas* at Walpi and
Mishongnovi. The Hopis freely cooperated with the missionaries at first. The missionaries' intent was, in the words of anthropologist E. Charles Adams,

> to convert the Pueblo villages to economically self-sufficient
> units that would participate in a regionwide economic system
> administered from the new seat of government in Santa Fe. To
> promote economic self-sufficiency in each village, the Francis-
> cans introduced domestic plants and animals and organized
> the people into work parties that were to build the mission
> and related buildings, plant and tend gardens and generally
> transform each village so that it would be able to support a
> mission with no outside help. (Adams 1989:82)

In this they were ultimately unsuccessful, for the Hopis were
able to accomplish what other groups could not after initial contact with the Spaniards: maintain the freedom to pick and choose
from among Spanish ideas and artifacts with no compulsion to
adopt them. Hopis chose to use domesticated animals, agricultural implements, and new types of food. In addition, their craft
traditions were enriched in the areas of weaving, metallurgy,
and woodworking. The Hopis rejected Spanish subsistence, settlement, and economic patterns, although a few Hopis adopted

Christianity and learned to read and write. While the Spaniards did occasionally try to enforce their views in a limited way, there is no evidence of raids on kivas to destroy religious objects, and missionaries who used harsh methods were removed. In addition, the number of missionaries was so small that their influence was almost inconsequential except at Awatovi. Nevertheless, conservative Hopi religious leaders resented the Spaniards' interference with religious practices and beliefs. Distrust and unfriendliness toward the intruders intensified in the face of the Hopis' determined and highly conscious choice to continue to live as Hopis.

Meanwhile, the Spaniards continued to encroach on Pueblo lands and to forcibly alter their way of life. In 1680 the Pueblo peoples and their allies united in revolt and drove the invaders from the province of New Mexico. Hopis killed five friars and demolished the churches and visitas but did not participate in the siege of Santa Fe. They enjoyed twelve years of freedom until the Spaniards returned in 1692, under Diego de Vargas, and reconquered the northern Rio Grande region. Vargas journeyed to Hopi and found the Hopis ready to resist, assisted by their Ute, Havasupai and Navajo allies. He gained nominal allegiance from some leaders (probably from Awatovi's pro-Spanish faction), but the Hopis were simply pretending to acquiesce to Spanish demands in order to avoid conflict. As soon as the Spaniards left, they went about their lives as they saw fit. In effect, Spanish reconquest meant little to the Hopis. Their isolation and the aridity of the region did not attract Spanish colonists, and the Hopis were able to go their own way.

Spanish influence did have three major effects. First, the Hopis moved some of their settlements in expectation of Spanish retribution. Shungopavi and Mishongnovi built new pueblos atop Second Mesa in 1680, and Walpi moved to the remote tip of First Mesa soon after. Sichomovi was built in the 1700s by people from Walpi and immigrants of Zuni's Asa clan, who were opposed to Spanish rule. A new pueblo, Sipaulovi, was founded in an inaccessible part of Second Mesa by families from Shungopavi in order to keep ritual equipment from falling into Spanish hands.

Second, the Hopis began to split into pro-Spanish and anti-

Spanish factions. In 1699 several Hopis from Awatovi went to Zuni or Santa Fe to request that missionaries rebuild the mission. The following year a group of Hopis led by Oraibi leader Francisco de Espeleta went to Santa Fe and requested that the mission be removed and that a peace treaty be signed that would ensure religious freedom for the Hopis. The Spanish governor, Pedro Rodríguez Cubero, refused, and the Hopis immediately attacked Awatovi in an effort to "eliminate all vestiges of the religion that had been imposed on them and the people who had introduced it" (Lomawaima 1989:96). The village was destroyed, all the men who resisted were killed, and the women and children were settled in other villages. Awatovi lands were distributed to the people of Walpi and Shungopavi. It is likely that all the Hopis who favored the return of the Spaniards were eliminated at one stroke, and anti-Spanish leaders became dominant everywhere.

While Spanish governors occasionally staged "punitive expeditions," the well-defended villages were never successfully stormed. The Hopis were more determined than ever to maintain their separate existence, which they did throughout the eighteenth century by isolating themselves and refusing to listen to outsiders who periodically visited the Hopi Mesas. In 1775, for example, elders directly intervened or publicly debated the Franciscans when the friars attempted to preach. Twice, however, the Hopis requested Spanish help. In the late 1770s and early 1780s, following several years of drought, crop failure, and smallpox, they asked the Spaniards to send food. When this was not forthcoming, most Hopis went to live in villages on the Rio Grande, at Zuni and Sandia, among the Havasupai, or with the Navajo at Canyon de Chelly, leaving only 800 people in the Hopi villages. The second request came after six years of increasingly devastating Navajo and Ute raids. In 1818 the Hopis asked the governor to send troops to fight the Navajos, but the Spaniards were unable to do so. Thus, except for these extraordinary situations, the Hopis made a clear choice to follow Hopivotskwani, not the Spanish way of life.

The third result of Spanish influence was that the Hopi Mesas became a refuge for those Pueblo people who could not bear Spanish domination. Several Rio Grande Pueblo groups—Tewa, southern Tiwa, and Keresan—moved to Hopi between 1681 and 1700. Tewa traditions state that they came at the invitation of Walpi elders who traveled to Tewageh (Tsawadeh) four times. The people of Walpi had been suffering from Ute attacks, and the elders asked the Tewas, whose bravery was widely known, to protect them. Tewas say they accepted the invitation as a humane service and built Hano (Tewa Village) on First Mesa and Payupki on Second Mesa to serve as guard posts at the head of the trails to Hopi. In return, they were given land and various privileges and duties.

Hano: Welcoming the Tewas

In 1775 Silvestre Vélez de Escalante noted that while Payupki no longer existed, Hano was flourishing. The people of Hano were tied to the Hopi social and ceremonial system in complex ways yet retained some independence. When the Tewas were invited to live at Hano, they gave heed to their town chief's counsel and came only as the summer moiety (leaving the winter moiety behind) and actively tried to remain a distinct group. The Tewas adopted Kachina ceremonies and other rituals but continued to speak their native language. Today, Tewas and Hopis belong to the same secret religious fraternities and share responsibilities in the Kachina ceremonies. A Tewa clown organization exists, and the Tewas place great emphasis on curing. All religious matters are controlled by a hierarchy headed by a village chief and a war chief responsible for guarding the pueblo against enemies and for settling internal quarrels. In effect, the Tewas' patrilineal moiety system (in which the village is divided into two parts for their social and ceremonial organization) has been superimposed on the Hopi female-based clan organization through extensive intermarriage. Nevertheless, matrilocal residence ensured that Tewa remained the language spoken at home. This is a strategy that anthropologist Edward Dozier feels is understandable "for a people in a social and numerically minority status" (1966:25).

Over the years it has been common for Hano, and through it the people of First Mesa, to experiment with European ideas that could enhance the Hopi Way. During the 1890s, for example, the Hopis adopted silver jewelry making from the Zunis, and Walpi

quickly became the center of production. Meanwhile, the famous pottery revival of Nampeyo occurred in Hano (see pl. 53).
People from Hano also were the first to settle at the mesa's base
and to send their children to school. They cooperated with government agents and tried new agricultural tools and new economic enterprises. They adopted new architectural features,
such as overhanging *vigas,* which they had seen at Zuni. As the
guards for Hopi society, such experimentation is an appropriate
activity for these people, and, given the next period of Hopi
history, one that kept the Tewa busy.

Increasing Contacts with the Paahana World

Unlike the Eastern Pueblos, the Hopis experienced more limited
contacts with whites during the late eighteenth and early nineteenth centuries. Raiding by Navajos, Utes, and Apaches halted
most travel in northern Arizona, although the Hopis continued
to trade with other Indians, especially the Zunis and Havasupais.
This does not mean that life was without incident, however, for
slave raiding did occur, with devastating effects on families.
Memories of these incidents were vivid years later. Early in the
twentieth century, Dawaventsie, the oldest woman in Walpi's
Water clan, recalled how she had barely escaped one such encounter:

> It was early one morning and I had gone with two other
> girls, cousins of mine, down to the spring at the foot of the
> mesa for water. These Mexican men came toward us, and we
> ran, but they caught up with us and started to take us away. I
> fought the man who was holding me and got loose and ran up
> the mesa trail faster than he could run.
>
> I rolled rocks on them when they tried to come up so they
> gave it up. I ran on up to the top of the mesa and gave the
> alarm and one man went to rescue the other two girls, but the
> Mexicans had horses and they got away with the girls, who
> have never been heard of to this day. (Lockett 1933:84–85)

Hopis and Tewas continued to be wary of the outside world even
when the outsiders came peacefully.

The third quarter of the nineteenth century brought new contacts as Anglo Americans began visiting Hopi in steadily increasing numbers—explorers, trappers, soldiers, surveyors, traders, missionaries, and eventually government agents, scholars, and tourists. These contacts were generally friendly, and the Hopis had a favorable impression of Anglo Americans, in contrast to their attitude toward Spaniards and Mexicans. In fact, some Hopis believed that the Anglos were a people identified in Hopi myth and prophesy as the Paahana (Bahana), who were to come from the east to help them. Evidence that the u.s. Army was trying to control raiders also contributed to the positive image. But the prophecy also said that the Hopis should take care. In 1885, Wicki, head of Walpi's Antelope Society, told Alexander Stephen that Spider Woman had foretold that there would be a time when the Paahana would come among the Hopis, and the Snake Society, having brave hearts, would be the first to make friends and learn good from them. But she warned, "the Hopitah are not to follow in the White man's footsteps but are to walk beside them, always keeping in the two footsteps of their fathers" (Stephen 1929).

The period of initial contact before 1900 was not always easy. Explorers found villages in 1853 ravaged by smallpox, and Lieutenant A. W. Whipple, leader of the u.s. Topographical Engineering Survey, reported that in one village he saw only the chief and one other man remaining of all the able-bodied men. Similar epidemics, along with a series of droughts and famines in the 1860s, cut the Hopi population dramatically, and raids meant that the people could not look far for food. As Dawaventsie of Walpi remembered, "Once food was so scarce that I had to go with my mother and sister to Second Mesa, and we stayed there with our clan relations till food was scarce, and then we went to Oraibi and stayed with our clan relations. This was a two-year famine and almost everybody left Walpi and wandered from village to village, living wherever they could get food" (Lockett 1933:85–86). But with the cessation of intertribal hostilities, Hopis began to move off the mesas to ranches, farms, and the new towns of Polacca and Keams Canyon on the lowlands south of the mesas. In 1870, Chief Tuba founded a farming settlement along Moencopi Wash near the home of the Havasupais, Paiutes, and Navajos. Mormons, who had become interested in the Hopis by 1858, hoped to produce woolen textiles and founded a short-lived mill in 1879. They also attempted to convert the Hopis and other Indians because they considered them Lamanites, one of the lost tribes of Israel. The Mormons were only mildly successful.

Unfortunately, old European patterns resurfaced as other missionary groups tried to convert the Hopis. In 1870, Protestants established a mission school at Keams Canyon, and Moravians settled at Oraibi. In 1875, Baptists established a mission at Mishongnovi. According to the Hopis, one of the most infamous conversion attempts was that of the Mennonites. Without village approval but with the consent of government officials, Mennonites sent Henry R. Voth to Oraibi in 1893. After learning the language, Voth, over elders' objections, attended kiva ceremonies. His extensive, detailed, and careful reports were published, to the horror of the Hopis, who considered this information secret. Like other attempts at conversion, the results were generally negative. More problematic were the missionaries' attempts to have Kachina dances stopped because they were "horrifying and unclean." The missionaries misunderstood the nature of the ceremonies and the antics of ritual clowns, called mudheads, who were vulgar and often sexually explicit when viewed through the lens of Victorian moral standards. Many tried to have government officials stop all Hopi public rituals.

The isolation of the Hopi villages from Anglo American settlements meant that the federal government interfered in their affairs later than it did with other groups. But with the completion of the railroad in 1881, government officials began to fear for Hopi land. In 1882, some 2,428,000 acres were federally acknowledged as the Hopi Reservation and distinguished from Navajo lands, which surrounded Hopi land, even though no treaty had been signed. The reservation, however, was not designated as land exclusively for the Hopis but was reserved for whatever tribes the secretary of the interior might desire to move there. This was an unfortunate inclusion that still causes legal problems to this day. Hopis had no knowledge of the new boundaries (or the theoretical opening of their area to other Indians), so of course they used more than the designated area. By the 1890s they ran livestock as far east as Ganado and as far northwest as Shonto.

While the Hopis were at first complacent and self-confident in their relationship with the government, outside contacts again began to present problems: Hopis split, as they had earlier, into pro- and anti-Anglo factions, usually identified as progressive and conservative. One instance of this process occurred during the dispute over sending children to Bureau of Indian Affairs (BIA) schools in the 1880s and 1890s. When the BIA school opened at Keams Canyon, Hano parents sent children immediately, but Oraibi refused to comply with government mandates and quotas. The headman, Lololoma, who had traveled to Washington in 1875 and who was impressed with Anglo culture, favored schooling. He was opposed, however, by Lomahongyoma, head of the Spider clan, who professed to believe that Lololoma's leadership was in conflict with Hopi sacred traditions. The village slowly split into two determined groups, and the fission escalated when government officials attempted to survey the villages in accordance with the Land Allotment Act of 1887. While all Hopi villages were opposed to individual allotment, troops were called out at Oraibi to counter promised physical resistance and the threatened burning of the schools. Unfortunately, the federal government overreacted and imprisoned five leaders. This act cemented the conservatives against the government. While the Hopis were eventually successful in eliminating the threat of allotment, they gradually acquiesced to government pressure and sent their children to school when day schools were placed in Oraibi, Toreva, and Polacca in 1897.

Third Mesa: New Villages Are Formed

In a society made up of many clans, some more powerful than others, the potential for conflict and rivalry was always present. The agricultural cycle and the complex religious calendar of the Hopis integrate the ceremonies of these various clans so that all clans have to work together to maintain order in the universe and to ensure rain, fertility, and the growth of crops. Marriage also cements cross-clan ties. In addition, gossip, criticism, and ridicule are major means of group social control. These factors tend to keep individuals from deviating from what the group considers to be normal and proper behavior. Hopis, in the past and today, are keenly sensitive to being singled out for public disapproval, laughter, or ostracism. If one does wrong, justice is imminent and observable; for example, illness may befall the family's village, a disastrous flood may destroy property, or severe drought may lay waste to an individual's crops. To divulge secrets can also bring disaster, as speech is restricted in certain circumstances and Hopis have a profound conviction about the reasonableness of secrecy. As long as one remains silent, "his power is still in him" (Parsons 1939:43). Thus there is much esoteric information about their culture that Hopis feel should not be divulged. As Don Talayesva of Oraibi remarked in his autobiography about his experiences with anthropologist Leo Simmons in 1938, "When I knew him better, I agreed to tell him anything about myself except ceremonial secrets" (Talayesva 1942:328). Individuals can be ostracized or accused of witchcraft for sharing such information with Hopis and non-Hopis who have not become members of the proper society or who were not born to a particular clan. Innovation is also suspect because the thing or idea has not been tried, and no Hopi wants to be the one to say yes to anything new lest it prove to be a bad idea. It is much safer to do nothing or to follow only in the footsteps of the ancestors in doing what has already been proven to be Hopivot-skwani.

Given this system of social control, when the system of clan checks and balances becomes unbalanced due to political, social, or economic tensions, a village tends to split and a new village is formed. This is an extremely old process. In the historical period it occurred when pressures to acculturate were laid on top of traditional reasons for disagreement and population pressures. On Third Mesa such a situation occurred in the 1890s and early 1900s. Oraibi, one of the largest Hopi villages, split into Hotevilla, Bacavi, and Kykotsmovi (New Oraibi).

Much has been written about the Oraibi split. Over a ten-year period, factionalism had developed as a result of disputes over the legitimate control of certain ceremonies and clan lands. Lololoma and his rivals opposed one another on religious matters and over the acceptance of innovations introduced or required by Anglo Americans. Lomahongyoma appeared to be questioning Lololoma's legitimate claim to the chieftainship at all times. While Oraibi was the most "conservative" Hopi village, some people there sent their children to government schools. These individuals were called "friendlies" or "progressives" and tended to be followers of Lololoma. The followers of Lomahongyoma were called "hostiles." These names are misleading, however, since the basic issues centered ultimately on disputes over Hopi ceremonial propriety and prerogative, land allocations, population pressures, traditional prophecy (*diingavi*), economic hardship, and power. Each household chose sides rather than adhering to old alignments by clan membership. By 1898, Lomahongyoma and his chief supporter, Yokioma, were defying Lololoma's authority in open meetings, and the two factions began to give ceremonies simultaneously. There was no cooperation, which is required by Hopivotskwani. Village life was seriously disrupted.

In 1904, when Lololoma was succeeded by Tewaquaptewa, the hostiles invited a group from Shungopavi to move to Oraibi. As village chief, Tewaquaptewa opposed this because rain had been scarce and there was not enough food to feed both residents and newcomers. Yokioma insisted and the Shungopavi people stayed on even though the friendlies sought to remove them. On September 8, 1906, the difficulties led to a final schism and violence. A formal pushing contest was held. One of the hostile leaders marked a line between the two groups and declared that whichever side could push him over the line would stay in the village. The friendlies won, and more than 300 hostiles and immigrants moved seven miles northwest on the mesa and founded Hotevilla. While the hostiles could have sought refuge at other villages, they did not, because village identity and loyalty were so important to the Hopis. They preferred to found a new village and establish a new ceremonial cycle, and thereby ensure political autonomy. The 324 friendlies remained in Oraibi under Tewaquaptewa's leadership.

Unexpectedly, the BIA agent appeared and arrested the hostile leaders and sent them to jail for a year. This left many Hotevilla families without men and caused economic hardship. Simultaneously, the agent announced that Tewaquaptewa and his assistants would have to go to school "to learn civilized ways." When these men returned from several years at a boarding school in Riverside, California, they were strongly anti-government and anti-Christian and refused to let anyone live in Oraibi who had shown an inclination toward Christianity. Several families moved to the foot of the mesa and founded Kykotsmovi in 1910 near Lorenzo Hubbell's trading post and the government school. Tewaquaptewa had earlier refused to allow a group to return from Hotevilla following arguments over BIA land allotments in 1907, and these families in turn founded Bacavi. Thus conservatism with regard to political and religious views, the pressure of a water shortage, Navajo encroachment on grazing land, and the Paahana led to the proliferation of settlements on Third Mesa. And it was diingavi, for Hopi prophesy had predetermined that villages would split when necessary.

The Indian Reorganization Act and Hopi Politics

It was to be expected that the Hopi Way and the American Way would occasionally be at odds. The federal government had treated the Hopis as a single tribe and political entity, not recognizing village autonomy or clan hierarchies. In the 1930s, BIA commissioner John Collier requested that the Hopis create one democratically elected government under the Indian Reorganization Act to oversee all the villages as a single unit and serve as a corporate entity to conduct business. The Hopi institution that came closest to being a legislative body was the annual elders' meeting for "Chief's Talk," held at the close of the Soyala ceremonies. Within each village, a theocracy, not a democracy, governed. Clan hierarchies followed the sequence of arrival: the first to arrive had the highest social status and power. Leadership was associated with religious positions; priests defined the annual calendar of events and therefore daily life. The heads of the ceremonial societies and clans constituted an assembly, with the priest of the most important society or clan serving as the *kikmongvi*, or village chief. Responsible for the welfare of the community, he selected his successor from among the sons of his female siblings, but he had limited power. Decisions were made by consensus following much discussion. This form of government often stood in stark contrast to American ideas of a representative, democratically elected government and the separation of church and state.

The time-consuming deliberations of Hopi elders continually frustrated government agents, who wanted quick responses to their proposals. Elders often refused to negotiate and compromise or to follow majority rule. In 1935, BIA officials sent Oliver La Farge to confer with the Hopis and develop a new pan-village government. La Farge presented a proposal for a tribal organization that was to consist of a tribal council with representatives from nine village units (First Mesa decided to act as a single unit) and to be composed of members who were at least twenty-five years of age, spoke Hopi or Tewa, and had lived on the reservation for at least two years. After certification by the respective village leaders, the representatives were to be certified by the outgoing council each year on December 1. The council was to elect a chairman and a vice-chairman from among its members and to appoint a secretary and a treasurer (who might or might not be council members), and two interpreters and a sergeant at arms (who were not council members).

When this constitution and by-laws, which advocated majority rule, were submitted for Hopi ratification on October 24, 1936, they were adopted by a vote of 651 to 104. In the election, however, only 30 percent of those entitled to vote cast ballots. In 1937 the first tribal council met, but individuals from Oraibi and Shungopavi refused to recognize its authority. Active members were nearly all from First Mesa villages and Kykotsmovi, and instead of unifying the Hopis, the new system further polarized the people. Nevertheless, the council, after a period of inactivity, was recognized as the official body of the "Hopi Tribe" in 1953, and members from all villages became active members.

The tribal council represented a progressive faction in the 1950s and 1960s and has continued to be viewed by some Hopis as an Anglo imposition on conservatives or traditionalists, as a body with which the federal government could deal, whether the "old ones" approved or not. The tribal council was opposed through nonparticipation; some members of the theocracy insisted on the autonomy of village deliberation and action and did not recognize the right of the BIA to concern itself with their affairs. In addition, these men contended that the sacred traditions of their identification with the land would be infringed upon by most tribal council actions. The council, as a result, dealt chiefly with BIA matters. It regulated tribal funds and commercial enterprises, maintained law and order, and protected Hopi arts, industries, and ceremonies. More recently, the council has represented Hopi interests with corporations, like Peabody Coal, and other American Indian governments, such as dealing with the Navajos in the Hopi-Navajo land dispute. Several members hold ceremonial leadership positions. Progressives and some conservatives have found it a very effective body from which to articulate Hopi voices to the outside world.

The Hopi Road Is Long, Strong, and Endless

For centuries the Hopi people have lived on the high plateau of northeastern Arizona, an area that serves as a natural protection from enemies. Here they developed a complex way of life. Several of their villages and their antecedent settlements, perched high atop three fingerlike mesas, the Hopi Mesas, are among the oldest continuously occupied communities in the United States. This continuity has been possible because the area around the mesas provides extensive and diverse resources—water from springs, arable land, plants, animals, and mineral resources. Before the coming of the Spaniards, the Hopis were skilled farmers, meeting the challenge of the region's arid climate, high elevations, and short growing season by developing specially adapted plants (corn, beans, squash, and tobacco) and by using small, well-situated plots. The Hopis were also skilled artists, traders, and miners of coal and turquoise who were successfully following Hopivotskwani.

Religion, art, and agriculture continue to play important roles in Hopi culture. Amid the modern realities of intertribal land disputes, unemployment, internal political disagreement, and off-reservation jobs, corn and Kachina ceremonies remain a central part of Hopi identity. This does not mean that there have not been changes. Agriculture is now primarily symbolic and cere-

A young girl in a home in the village of Sichomovi in 1902. Hopi contacts with outsiders brought many new items of material culture, such as metal containers and umbrellas, yet Hopi aesthetics, as seen in the house decorations shown here, remained strong. (Photograph by A. C. Vroman, Los Angeles County Museum of Natural History)

monial—corn is grown for rituals rather than as a subsistence food. There have been some modifications in the rich cultural heritage to meet the demands of modern life. Many ceremonies are scheduled for weekends, for example, so that off-reservation workers can return and participate. Due to demographic shifts resulting from two centuries of epidemics, some clans no longer have enough people to perform certain rituals, and membership in some of the religious societies is waning. In the modern villages, some religious societies were never established, so individuals return to the older villages to take part in rites. Many young people compensate for this by eagerly participating in the flourishing Kachina ceremonies, thereby confirming their ties to the prehistoric "footprints" left by their ancestors. Kivas are being enlarged to accommodate the growing number of participants.

Hopis are very concerned with education. In the past, formal education was viewed in the same light as missionization, as an imposition and a threat to Hopi values. When it became clear that compulsory education was not going away and could have value if done with dignity, the Hopis made a conscious decision to accept schooling. Not learning English came to be seen as an unwitting way of giving the white man power over Hopi lives. Today Hopis value education not to become like Anglo Americans but to take advantage of the opportunity to chose what is best for individuals, families, and the group, and to reject what is potentially harmful to the Hopi way of life. Education is now an integral part of Hopi community life, and it includes formal classes in the Hopi language.

As a result of this perspective, local educational facilities have been upgraded in recent years and provide both elementary and secondary education. Many students are enrolled in college, and the numbers are increasing. There are still problems that must be worked out, however. According to Emory Sekaquaptewa (1988) since each community feels it has a slightly different dialect, the villages cannot decide which should be formerly taught in the Hopi high school as the "official" Hopi language. No one village wants to impose its dialect on the others. The same problem arises as the Hopis construct a Hopi-English dictionary.

Hopivotskwani is still the core with respect to which Hopi life evolves. For most Hopis, it is an important motivating factor in daily life. Hopis are religious people, and most adhere to traditional religious beliefs despite limited inroads by Christianity in some villages. After many battles, Hopis have gained legal recognition that their religion is equal to other religions in the United States. From fighting against BIA bans on clowning activities through their protests against imitation and the use of sacred masks and costumes in secular contexts, Hopis have been successful. Since the 1880s, Hopis have objected to the use of copies of sacred paraphernalia in Anglo secular celebrations, most recently by "New Agers," and during the 1980s they obtained apologies from the organizations involved. The most recent has been the Smokis in Prescott, a secret group of Anglo businessmen who for years staged "Snake Dances." Following Hopi pickets, the Smokis stopped their August performances.

Contemporary life styles have brought changes to all Hopi villages. Jobs have drawn Hopis away from their land. By 1980 a quarter of the tribe had dispersed across the Southwest, working as college professors, lawyers, artists, engineers, architects, carpenters, police officers, forest fighters, ranchers, schoolteachers, entrepreneurs, or mechanics. Yet all return to fulfill their responsibilities to their clan, and all consider the Hopi Mesas their home. There is no question that the Hopi Way continues to exist in the 1990s, against a background of life in a heterogeneous world. Hopis are both members of a group and individuals, and those who follow Hopivotskwani are, in the words of Edward Spicer,

> militant Baptists to rigid followers of the Old Oraibi "traditional" leadership. They range from college graduates aware of most of the elements of Anglo culture, to aged men and women who have the barest acquaintance with only a few material items of Anglo culture. They range from Hopi men married to vigorously progressive Tewa women, to Hopi girls married to Filipino migrants to the United States. They range from families still living on clan land and following through all the traditional ceremonies of their village, to families paying rent in Los Angeles on modern apartments and returning to their villages for an occasional Niman Kachina or Soyal ceremony. They range from men desperately wanting to obtain cash for land settlements or oil leases, to men convinced that the ancient sacred relationship to the land has nothing to do with cash. (Spicer 1962:208–9)

But all are Hopis, for "the fabric of Pueblo culture was and continues to be tough and flexible, giving wherever and whenever necessary, but scarcely tearing, much less shredding" (Lomawaima 1989:97). Hopis have adopted, and will continue to test, various facets of life that come from the outside, weaving some features into the Hopi fabric of life and rejecting those that do not meet the test of Hopivotskwani.

REFERENCES AND
SUGGESTED READINGS

Adair, John

1944   *The Navajo and Pueblo Silversmiths.* Norman: University of
        Oklahoma Press.

Adams, E. Charles

1989   Passive Resistance: Hopi Responses to Spanish Contact and
        Conquest. In *Columbian Consequences,* vol. 1: *Archaeologi-
        cal and Historical Perspectives on the Spanish Borderlands
        West,* edited by David H. Thomas, 77–92. Washington,
        D.C.: Smithsonian Institution Press.

1991   *The Origin and Development of the Pueblo Katsina Cult.* Tuc-
        son: University of Arizona Press.

Bahr, Donald

1975   *Pima and Papago Ritual Oratory.* San Francisco: Indian His-
        torian Press.

Bahr, Donald, Juan Gregorio, David Lopez, and Albert Alvarez

1974   *Piman Shamanism and Staying Sickness (Ká:cim Múm-
        kidag).* Tucson: University of Arizona Press.

Bahr, Donald, Juan Smith, William Smith Allison, and Julian Hayden

1994   *The Short, Swift Time of Gods on Earth: The Hohokam
        Chronicles.* Berkeley: University of California Press.

Bahre, Conrad

1967   The Reduction of Seri Range and Residence in the State of
        Sonora, Mexico (1536–Present). M.A. thesis, University of
        Arizona.

1980   Historic Seri Residence, Range, and Socio-Political Struc-
        ture. *The Kiva* 45:197–209.

Balke, Bruno, and Clyde Snow

1965   Anthropological and Physiological Observations on Tara-
        humara Endurance Runners. *American Journal of Physical
        Anthropology* 23:293–301.

Barnett, Franklin
1968   *Viola Jimulla: The Indian Chieftess.* Yuma, Ariz.: Southwest Printers.

Basso, Keith
1969   *Western Apache Witchcraft.* Anthropological Papers of the University of Arizona 15. Tucson: University of Arizona Press.
1970   *The Cibecue Apache.* New York: Holt, Rinehart and Winston.
1971   *Western Apache Raiding and Warfare, from the Notes of Grenville Goodwin.* Tucson: University of Arizona Press.
1987   "Stalking with Stories": Names, Places and Moral Narratives among the Western Apache. In *On Nature,* edited by Daniel Halpern, 95–116. San Francisco: North Point Press.

Beaglehole, Earnest, and Pearl Beaglehole
1935   *Hopi of the Second Mesa.* Memoirs of the American Anthropological Association 44. Menasha, Wis.

Begay, Keats, et al.
1977   *Navajos and World War II.* Tsaile, Ariz.: Navajo Community College.

Bennett, Wendell, and Robert Zingg
1935   *The Tarahumara: An Indian Tribe of Northern Mexico.* Chicago: University of Chicago Press.

Bingham, Sam, and Janet Bingham, eds.
1984   *Between Sacred Mountains: Navajo Stories and Lessons from the Land.* Tucson: University of Arizona Press.

Bishop, Ronald L., Valetta Canouts, Suzanne P. De Atley, Alfred Qoyawayma, and C. W. Aikens
1988   The Formation of Ceramic Analytical Groups: Hopi Pottery Production and Exchange, A.D. 1300–1600. *Journal of Field Archaeology* 15:317–37.

Bourke, John G.
1891   *On the Border with Crook.* Reprint, Lincoln: University of Nebraska Press, 1971.

Bowen, Thomas
1983   Seri. In Ortiz 1983:230–49.

Brown, F. Lee, and Helen Ingram
1987   *Water and Poverty in the Southwest.* Tucson: University of Arizona Press.

Brown-Kampden, Catherine
1978   The Maze of Life Design of the Pima-Papago. *The Masterkey* 52 (2): 67–70.

Bunte, Pamela A., and Robert J. Franklin
    1987   *From the Sands to the Mountain: Change and Persistence in a Southern Paiute Community*. Lincoln: University of Nebraska Press.

Buskirk, Winfred
    1986   *The Western Apache: Living with the Land before 1950*. Norman: University of Oklahoma Press.

Carmony, Neil, and David Brown
    1983   *Tales from Tiburon: An Anthology of Adventures in Seriland*. Phoenix: Southwest Natural History Association.

Castetter, Edward, and Willis Bell
    1942   *Pima and Papago Indian Agriculture*. Albuquerque: University of New Mexico Press.

Coffeen, William R.
    1972   The Effects of the Central Arizona Project on the Fort McDowell Indian Community. *Ethnohistory* 19:345–77.

Corbusier, William H.
    1969   *Verde to San Carlos: Reflections of a Famous Army Surgeon and His Observant Family on the Western Frontier, 1869–1886*. Tucson: Dale Stuart King.

Correll, J. Lee
    1979   *Through White Men's Eyes: A Contribution to Navajo History*. Vol. 6. Window Rock, Ariz.: Navajo Heritage Center.

Coues, Elliot
    1900   *On the Trail of a Spanish Pioneer: The Diary and Itinerary of Francisco Garcés*. 2 vols. New York: Francis B. Harper.

Courlander, Harold S.
    1971   *The Fourth World of the Hopis*. New York: Crown Publishers.
    1982   *Hopi Voices: Recollections, Traditions, and Narratives of the Hopi Indians*. Albuquerque: University of New Mexico Press.

Crumrine, N. Ross
    1977   *The Mayo Indians of Sonora: A People Who Refuse to Die*. Tucson: University of Arizona Press.

Deeds, Susan
    1989   Rural Work in Nueva Vizcaya: Forms of Labor Coercion on the Periphery. *Hispanic American Historical Review* 69:425–49.

DiPeso, Charles C., and Daniel S. Matson
    1965   The Seri Indians in 1692 as Described by Adam Gilg, S.J. *Arizona and the West* 7:33–56.

Dobyns, Henry F., and Robert C. Euler

1960   A Brief History of the Northeastern Pai. *Plateau* 32 (3): 19–56.

1970   *Wauba Yuma's People: The Comparative Socio-Political Structure of the Pai Indians of Arizona.* Prescott, Ariz.: Prescott College Press.

1971   *The Havasupai People.* Phoenix: Indian Tribal Series.

1976   *The Walapai People.* Phoenix: Indian Tribal Series.

Dobyns, Henry F., Paul H. Ezell, and Greta S. Ezell

1963   Death of a Society. *Ethnohistory* 10:105–61.

Dobyns, Henry F., Paul H. Ezell, Alden W. Jones, and Greta S. Ezell

1957   Thematic Changes in Yuman Warfare: Cultural Stability and Cultural Change. In *Proceedings of the American Ethnological Society,* 46–71. Seattle.

Dozier, Edward P.

1966   *Hano: A Tewa Indian Community in Arizona.* New York: Holt, Rinehart and Winston.

Dyk, Walter

1938   *Son of Old Man Hat: A Navaho Autobiography.* Lincoln: University of Nebraska Press.

Eggan, Fred

1950   *Social Organization of the Western Pueblos.* Chicago: University of Chicago Press.

Emory, William H.

1857   *Report on the United States and Mexican Boundary Survey.* U.S. Congress, 34th Cong., 1st sess., House Exec. Doc. no. 135.

Euler, Robert C.

1966   *Southern Paiute Ethnohistory.* University of Utah Anthropology Papers 78. Salt Lake City: University of Utah Press.

1972   *The Paiute People.* Phoenix: Indian Tribal Series.

Evers, Larry

1978   The Origin of the Crown Dance: An Apache Narrative with Rudolph Kane. In *Words & Place: Native Literature from the American Southwest.* Tucson: University of Arizona Radio-TV-Film Bureau.

Evers, Larry, and Felipe S. Molina

1987   *Yaqui Deer Songs / Maso Bwikam.* Tucson: University of Arizona Press.

1992   The Holy Dividing Line: Inscription and Resistance in Yaqui Culture. *Journal of the Southwest* 34:3–46.

Ezell, Paul

    1961   *The Hispanic Acculturation of the Gila River Pimas*. American Anthropological Association Memoir 90. Menasha, Wis.

    1983   History of the Pima. In Ortiz 1983:149–60.

Farnham, Thomas J.

    1843   *Travels in the Great Western Prairies, the Anahuac and Rocky Mountains, and in the Oregon Territory*. Reprint, *Early Western Travels*. Vols. 28–29. Edited by Reuben Gold Thwaites. Cleveland, Ohio: Arthur H. Clarke Co., 1906.

Felger, Richard, and Mary Beck Moser

    1985   *People of the Desert and Sea: Ethnobotany of the Seri Indians*. Tucson: University of Arizona Press.

Ferg, Alan, ed.

    1987   *Western Apache Material Culture: The Goodwin and Guenther Collections*. Tucson: University of Arizona Press.

Fontana, Bernard L.

    1958   History of the Colorado River Reservation. In *Social and Economic Studies: Colorado River Reservation*. Tucson: Bureau of Ethnic Research, Department of Anthropology, University of Arizona.

    1963   The Hopi-Navajo Colony on the Lower Colorado River. *Ethnohistory* 10:163–82.

    1971   The Seri Indians in Perspective. Introduction to *The Seri Indians*, by W. J. McGee. Reprint, Glorieta, N.M.: Rio Grande Press.

    1979   *Tarahumara: Where Night Is the Day of the Moon*. Photographs by John P. Schaefer. Flagstaff, Ariz.: Northland Press.

    1981   *Of Earth and Little Rain: The Papago Indians*. Reprint, Tucson: University of Arizona Press, 1989.

    1983a   Pima and Papago: Introduction. In Ortiz 1983:125–36.

    1983b   History of the Papago. In Ortiz 1983:137–48.

    1994   *Entrada: The Legacy of Spain & Mexico in the United States*. Tucson: Southwest Parks and Monuments Association.

Fontana, Bernard, Edmond Faubert, and Barney Burns

    1977   *The Other Southwest: Indian Arts and Crafts of Northwestern Mexico*. Phoenix: Heard Museum.

Forbes, Jack D.

    1965   *Warriors of the Colorado: The Yumas of the Quechan Nation and Their Neighbors*. Norman: University of Oklahoma Press.

Fowler, Catherine S., and Don D. Fowler

1981   The Southern Paiute, A.D. 1400–1776. In *The Protohistoric Period in the North American Southwest, A.D. 1450–1700*, edited by David R. Wilcox and W. Bruce Masse, 129–62. Arizona State University Anthropological Research Papers 24. Tempe: Arizona State University.

Freire-Marreco, Barbara

1914   Tewa Kinship Terms from the Pueblo of Hano, Arizona. *American Anthropologist* 16:268–89.

Gaillard, D. D.

1894   The Papago of Arizona and Sonora. *American Anthropologist* 7:293.

1896   The Perils and Wonders of a True Desert. *Cosmopolitan* 21:592–605.

Getty, Harry T.

1963   *The San Carlos Indian Cattle Industry*. Anthropological Papers of the University of Arizona 7. Tucson: University of Arizona Press.

Gifford, E. W.

1932   The Southeastern Yavapai. *University of California Publications in American Archaeology and Ethnology* 29:177–252.

1936   Northeastern and Western Yavapai. *University of California Publications in American Archaeology and Ethnology* 34: 247–354.

Goodwin, Grenville

1969   *The Social Organization of the Western Apache*. Tucson: University of Arizona Press.

Griffen, William

1959   *Notes on Seri Culture, Sonora, Mexico*. Monographs of the School of Inter-American Studies 10. Gainesville: University of Florida Press.

Griffith, James

1983   Kachinas and Masking. In Ortiz 1983:764–77.

Hackenberg, Robert

1983   Pima and Papago Ecological Adaptations. In Ortiz 1983: 161–77.

Hardy, R. W. H.

1829   *Travels in the Interior of Mexico in 1825, 1826, 1827, and 1828*. Reprint, Glorietta, N.M.: Rio Grande Press, 1977.

Hayden, Julian D.

1987   The Vikita Ceremony of the Papago. *Journal of the Southwest* 29:273–324.

Hayes, Alden C., and Karen Hayes

   1991   Sierra Madre Revisited. *Journal of Arizona History* 32:125–52.

Helliker, Kevin

   1988   Showdown in Parker. *Arizona Trend* 2 (6): 35–39.

Hinton, Leanne, and Lucille J. Watahomigie, eds.

   1984   *Spirit Mountain: An Anthology of Yuman Story and Song.* Tucson: University of Arizona Press.

Hough, Walter

   1915   *The Hopi.* Cedar Rapids, Ia.: Torch Press.

Howard, E. A.

   1887   Report of Agents in Arizona: Pima Agency, 1887. *Annual Report of the Commissioner for Indian Affairs to the Secretary of the Interior.* Washington, D.C.: U.S. Government Printing Office.

Hu-DeHart, Evelyn

   1984   *Yaqui Resistance and Survival: The Struggle for Land and Autonomy, 1821–1910.* Madison: University of Wisconsin Press.

Iverson, Peter

   1982   *Carlos Montezuma and the Changing World of American Indians.* Albuquerque: University of New Mexico Press.

James, Harry C.

   1976   *Pages from Hopi History.* Tucson: University of Arizona Press.

Joe, Jennie

   1988   Personal Communication. Exhibition Project Consultation. Tucson: Arizona State Museum.

Johnson, Broderick H., ed.

   1977   *Stories of Traditional Navajo Life and Culture by Twenty-two Navajo Men and Women.* Tsaile, Ariz.: Navajo Community College Press.

Johnston, Bernice

   1980   *The Seri Indians of Sonora, Mexico.* Tucson: University of Arizona Press.

Keller, Donald R., and Pat H. Stein

   1985   *Archaeological Study at Three Twentieth Century Yavapai Wickiup Sites, Prescott, Arizona.* Flagstaff: Museum of Northern Arizona.

Kelly, Isabel T.

   1939   Southern Paiute Shamanism. *Anthropological Records* 2: 151–67.

1964   *Southern Paiute Ethnography*. University of Utah Anthropological Papers 69. Salt Lake City: University of Utah Press.

Kelly, Isabel T., and Catherine S. Fowler
1986   Southern Paiute. In *Handbook of North American Indians*, vol. 11: *Great Basin*, edited by W. L. d'Azevedo, 368–97. Washington, D.C.: Smithsonian Institution Press.

Kelly, Lawrence C.
1968   *The Navajo Indians and Federal Indian Policy*. Tucson: University of Arizona Press.

1970   *Navajo Round Up: Selected Correspondence of Kit Carson's Expedition against the Navajo, 1863–1865*. Boulder, Colo.: Pruett Publishing Co.

Kennedy, John
1978   *Tarahumara of the Sierra Madre: Beer, Ecology, and Social Organization*. Arlington Heights, Ill.: AHM Publishing Corporation.

Kluckhohn, Clyde, W. W. Hill, and Lucy Wales Kluckhohn
1971   *Navaho Material Culture*. Cambridge, Mass.: Belknap Press of Harvard University Press.

Kluckhohn, Clyde, and Dorothea Leighton
1946   *The Navaho*. Reprint, Garden City, N.Y.: Doubleday, 1962.

Knack, Martha C.
1980   *Life Is With People: Household Organization of the Contemporary Southern Paiute Indians*. Socorro, N.M.: Ballena Press.

Kroeber, Alfred L.
1922   Earth-Tongue, a Mohave. In *American Indian Life*, edited by Elsie Clews Parsons, 189–202. New York: B. W. Huebsch.

1931   *The Seri*. Southwest Museum Paper 6. Los Angeles: Southwest Museum.

1948   Seven Mohave Myths. *University of California Anthropological Records* 11:1–70.

Kroeber, Alfred L., ed.
1935   *Walapai Ethnography*. American Anthropological Association Memoir 42. Menasha, Wis.

Kroeber, Clifton B., and Bernard L. Fontana
1986   *Massacre on the Gila: An Account of the Last Major Battle between American Indians, with Reflections on the Origin of War*. Tucson: University of Arizona Press.

Kroeber, Alfred L., and Clifton B. Kroeber
1973   *A Mohave War Reminiscence, 1854–1880*. Berkeley and Los Angeles: University of California Press.

Laird, Carobeth
1976   *The Chemehuevis*. Banning, Calif.: Malki Museum Press.
Lee, Richard
1972   The Kung Bushmen of Botswana. In *Hunters and Gatherers Today*, edited by M. G. Bicchieri, 326–68. New York: Holt, Rinehart and Winston.
Lincoln, Louise
1982   Navajo Silver, Navajo Aesthetics. In *Southwest Indian Silver from the Doneghy Collection*, edited by Louise Lincoln, 37–50. Austin: University of Texas Press.
Locke, Patrice
1984   Annie Dodge Wauneka: An American Original. *Indian Trader* 15 (12): 4–7.
Lockett, Hattie Greene
1933   *Unwritten Literature of the Hopi*. University of Arizona Social Science Bulletin 2. Tucson: University of Arizona.
Lomawaima, Hartman H.
1989   Hopification: A Strategy for Cultural Preservation. In *Columbian Consequences*, vol. 1: *Archaeological and Historical Perspectives on the Spanish Borderlands West*, edited by David H. Thomas, 93–100. Washington, D.C.: Smithsonian Institution Press.
Lowell, Edith
1970   A Comparison of Mexican and Seri Indian Versions of the Legend of Lola Casanova. *The Kiva* 35:144–58.
Lumholtz, Carl
1902   *Unknown Mexico*. New York: Charles Scribner's Sons.
Malotki, Ekkert, and Michael Lomatuway'ma
1987   *Stories of Maasaw, a Hopi God*. Lincoln: University of Nebraska Press.
Martin, John F.
1990   The Havasupai. *Plateau* 56 (4): 3–32.
McGee, W. J.
1898   The Seri Indians of Bahia Kino and Sonora, Mexico. *Seventeenth Annual Report of the Bureau of American Ethnology for 1895–1896*. Pt. 1:1–344. Washington, D.C.: Smithsonian Institution.
McGreevy, Susan B., and Andrew H. Whiteford
1986   *Translating Tradition: Basketry Arts of the San Juan Paiute*. Santa Fe: Wheelwright Museum of the American Indian.
McGuire, Thomas
1986   *Politics and Ethnicity on the Río Yaqui: Potam Revisited*. Tucson: University of Arizona Press.

1991   Indian Water Rights Settlements: A Case Study in the Rhetoric of Implementation. *American Indian Culture and Research Journal* 15:139–69.

McNitt, Frank
1963   *The Indian Traders.* Norman: University of Oklahoma Press.

Merrill, William
1983   God's Saviors in the Sierra Madre. *Natural History* 93 (3): 58–67.
1987   The Rarámuri Stereotype of Dreams. In *Dreaming: Anthropological and Psychological Interpretations,* edited by Barbara Tedlock, 194–219. Cambridge: Cambridge University Press.
1988   *Rarámuri Souls: Knowledge and Social Process in Northern Mexico.* Washington, D.C.: Smithsonian Institution Press.
n.d.   Rarámuri Easter. Manuscript in possession of the author.

Miller, Joseph
1952   *The Arizona Story.* New York: Hastings House.

Montgomery, Ross G., Watson Smith, and John Otis Brew
1949   *Franciscan Awatovi.* Papers of the Peabody Museum of American Archaeology and Ethnology 36. Cambridge, Mass.: Harvard University.

Moser, Edward
1963   Seri Bands. *The Kiva* 28 (3): 14–27.

Nabhan, Gary
1982   *The Desert Smells Like Rain: A Naturalist in Papago Country.* San Francisco: North Point Press.
1986a   Papago Indian Desert Agriculture and Water Control in the Sonoran Desert, 1697–1934. *Applied Geography* 6:43–59.
1986b   'Ak-ciñ 'arroyo mouth' and the Environmental Setting of the Papago Indian Fields in the Sonoran Desert. *Applied Geography* 6:61–75.

Ortega, Will
1994   Interview with Bobbi Clark. *Many Voices: Today's Native Americans.* Tucson: KUAT and KUAZ Radio, University of Arizona.

Ortiz, Alfonso, ed.
1983   *Handbook of North America Indians.* Vol. 10: *Southwest.* Washington, D.C.: Smithsonian Institution Press.

Pablo, Sally Giff
1983   Contemporary Pima. In Ortiz 1983:212–16.

Parman, Donald L.
1976   *The Navajos and the New Deal.* New Haven: Yale University Press.

Parsons, Elsie Clews
  1939   *Pueblo Indian Religion.* 2 vols. Chicago: University of Chicago Press.

Paul, Doris A.
  1972   *The Navajo Code Talkers.* Philadelphia: Dorrance and Company.

Pennington, Campbell
  1963   *The Tarahumara of Mexico: Their Environment and Material Culture.* Salt Lake City: University of Utah Press.
  1969   *The Tepehuan of Chihuahua: Their Material Culture.* Salt Lake City: University of Utah Press.
  1983   Tarahumara. In Ortiz 1983:306–14.

Powell, John Wesley
  1875   The Ancient Province of Tusayan. *Scribner's Magazine* 11: 193–213.

Powell, J., and G. W. Ingalls
  1874   Report on the Condition of the Ute Indians of Utah; the Pai-Utes of Utah, Northern Arizona, Southern Nevada, and Southeastern California; and the Western Shoshones of Nevada. In *Report on Commission of Indian Affairs for 1873*, 41–46. Washington, D.C.: U.S. Government Printing Office.
  1895   *Canyons of the Colorado.* Reprinted as *The Exploration of the Colorado River and Its Canyons.* New York: Dover, 1961.

Rea, Amadeo
  1983   *Once a River: Bird Life and Habitat Changes on the Middle Gila.* Tucson: University of Arizona Press.
  ms.   At the Desert's Green Edge: An Ethnobotany of the Gila River Pima. Partial ms. in possession of the editors.

Reichard, Gladys A.
  1950   *Navaho Religion: A Study of Symbolism.* New York: Pantheon Press.

Roberts, John M.
  1951   *Three Navaho Households.* Papers of the Peabody Museum of American Archaeology and Ethnology 40 (3). Cambridge, Mass.: Harvard University.

Roessel, Ruth, ed.
  1984   *Navajo Stories of the Long Walk Period.* Tsaile, Ariz.: Navajo Community College Press.
  1988   *Navajo Stories of the Long Walk Period.* Tsaile, Ariz.: Navajo Community College Press.

Russell, Frank
  1975   *The Pima Indians.* Tucson: University of Arizona Press.

Sauer, Carl
1934 The Distribution of Aboriginal Tribes and Languages in Northwestern Mexico. *Ibero-Americana* 5. Berkeley: University of California Press.

Saxton, Dean, and Lucille Saxton
1973 *Legends and Lore of the Papago and Pima Indians.* Tucson: University of Arizona Press.

Schroeder, Albert H.
1953 A Brief History of the Havasupai. *Plateau* 25 (2): 45–52.

Sekaquaptewa, Emory
1972 Preserving the Good Things of Hopi Life. In *Plural Society in the Southwest,* edited by Edward Spicer and Raymond H. Thompson, 239–260. New York: Interbook.
1988 Personal Communication. Exhibition Project Consultation. Tucson: Arizona State Museum.

Sheridan, Thomas
1979 Cross or Arrow? The Breakdown in Spanish-Seri Relations, 1729–1750. *Arizona and the West* 21:317–34.
1982 Seri Bands in Cross-Cultural Perspective. *The Kiva* 47:185–213.
1988 Kino's Unforeseen Legacy: The Material Consequences of Missionization among the Northern Piman Indians of Arizona and Sonora. *The Smoke Signal* 49–50. Tucson: Tucson Corral of Westerners.

Sheridan, Thomas, and Richard Felger
1977 Indian Utilization of Eelgrass (*Zostera marina L.*) in Northwestern Mexico: The Spanish Colonial Record. *The Kiva* 43:89–92.

Sheridan, Thomas, and Thomas Naylor
1979 *Rarámuri: A Tarahumara Colonial Chronicle.* Flagstaff, Ariz.: Northland Press.

Sine, Duke Wassaja
1988 Personal communication with Bruce Hilpert. Copy at the Arizona State Museum, Tucson.

Spicer, Edward H.
1962 *Cycles of Conquest: The Impact of Spain, Mexico, and the United States on the Indians of the Southwest, 1533–1960.* Tucson: University of Arizona Press.
1980 *The Yaquis: A Cultural History.* Tucson: University of Arizona Press.
1992 Excerpts from the "Preliminary Report on Potam." *Journal of the Southwest* 34:111–28.

Spicer, Edward H., and Raymond H. Thompson, eds.
1972 *Plural Society in the Southwest.* New York: Interbook.

Spier, Leslie

1928   Havasupai Ethnography. *Anthropological Papers of the American Museum of Natural History* 29:81–392.

1933   *Yuman Tribes of the Gila River.* Chicago: University of Chicago Press.

1936   *Cultural Relations of the Gila River and Lower Colorado Tribes.* New Haven: Yale University Press.

Stephen, Alexander M.

1929   Hopi Tales. *Journal of American Folklore* 42, no. 163: 1–72.

Stewart, Irene

1980   *A Voice in Her Tribe: A Navajo Woman's Own Story.* Socorro, N.M.: Ballena Press.

Stoffle, Richard, and Henry F. Dobyns

1983   *Niagantu.* Cultural Resource Series Monograph 7. Reno: Nevada State Office, Bureau of Land Management.

Talayesva, Don C.

1942   *Sun Chief: The Autobiography of a Hopi Indian.* Edited by Leo Simmons. New Haven: Yale University Press.

Titiev, Mischa

1941   *Old Oraibi: A Study of the Hopi Indians of Third Mesa.* Papers of the Peabody Museum of American Archaeology and Ethnology 22 (7). Reprint, Albuquerque: University of New Mexico Press, 1991.

Udall, Louise

1969   *Me and Mine: The Life Story of Helen Sekaquaptewa, As Told to Louise Udall.* Tucson: University of Arizona Press.

Underhill, Ruth

1936   *The Autobiography of a Papago Woman.* American Anthropological Association Memoir 46. Menasha, Wis.

1938   *Singing for Power: The Song Magic of the Papago Indians of Southern Arizona.* Reprint, Tucson: University of Arizona Press, 1993.

1939   *Social Organization of the Papago Indians.* New York: Columbia University Press.

1946   *Papago Indian Religion.* New York: Columbia University Press.

Underhill, Ruth, Donald Bahr, Baptisto López, José Pancho, and David Lopez

1979   *Rainhouse and Ocean: Speeches for the Papago Year.* Flagstaff: Museum of Northern Arizona Press.

U.S. Indian Claims Commission

1950   *The Hualapai and Yavapai Indians.* Report by Henry F. Dobyns et al. New York: Garland Publishing.

U.S. Senate

    1936   *Walapai Papers.* 74th Cong., 2d sess., Sen. Doc. no. 273.

Watahomigie, Lucille J.

    1984   The Hualapai. In *Spirit Mountain: An Anthology of Yuman Story and Song,* edited by Leanne Hinton and Lucille J. Watahomigie, 11. Tucson: University of Arizona Press.

Watahomigie, Lucille J., Jorigine Bender, and Akira Y. Yamamoto

    1982   *Hualapai Reference Grammar.* California: American Indian Studies Center, University of California at Los Angeles.

Webb, George

    1959   *A Pima Remembers.* Tucson: University of Arizona Press.

Whiteley, Peter

    1988   *Bacavi: Journey to Reed Springs.* Flagstaff, Ariz.: Northland Press.

Whiting, Alfred F.

    1985   *Havasupai Habitat: Ethnography of a Traditional Indian Culture.* Edited by Steven A. Weber and P. David Seaman. Tucson: University of Arizona Press.

Wilcox, David R., and W. Bruce Masse, eds.

    1981   *The Protohistoric Period in the North American Southwest,* A.D. 1450–1700. Arizona State University Anthropological Research Papers 24. Tempe: Arizona State University.

Williams, Kitty

    1994   Return of the Tarahumara. *Native Peoples* 7 (3): 20–27.

Wood, John J., Walter L. Vannete, and Michael J. Andrews

    1982   *"Sheep Is Life": An Assessment of Livestock Reduction in the Former Navajo-Hopi Joint Use Area.* Northern Arizona University Anthropological Paper 1. Flagstaff: Northern Arizona University.

Wyman, Leland C.

    1970   *Blessingway.* Tucson: University of Arizona Press.

Xavier, G. Harrington

    1941   Seri Indian Drawings. Ms. in the Arizona State Museum, Tucson.

Yava, Albert

    1978   *Big Falling Snow: A Hopi-Tewa Indian's Life and Times and the History and Traditions of His People.* Albuquerque: University of New Mexico Press.

Yetman, David

    1988   *Where the Desert Meets the Sea: A Trader in the Land of the Seri Indians.* Tucson: Pepper Publishing.

Zolbrod, Paul G.

    1984   *Diné Bahane: The Navajo Creation Story.* Albuquerque: University of New Mexico Press.

CONTRIBUTORS

E. Charles Adams received his Ph.D. in 1975 from the University of Colorado, Boulder. Since 1969 Adams has conducted archaeological fieldwork on the Four Corners area, specializing in the late prehistoric and historic Hopi occupation of northeastern Arizona. To that end, since 1985 Adams has directed the Homol'ovi Research Program at the Arizona State Museum. This program involves students and volunteers in the excavation and analysis of ancestral Hopi ruins (A.D. 1260–1400) in Homolovi Ruins State Park, along the Little Colorado River near Winslow. Adams has published extensively on his research, highlighted by his book *The Origin and Development of the Pueblo Katsina Cult* (University of Arizona Press, 1991).

Diane D. Dittemore is the curator of the ethnological collections at the Arizona State Museum. She conducted fieldwork with Seri Indians in Desemboque, Sonora, in 1978 and 1979, focusing on Seri one-stringed fiddles.

Catherine S. Fowler is a professor of anthropology at the University of Nevada, Reno. Dr. Fowler specializes in Great Basin ethnography, Numic and Uto-Aztecan linguistics, ethnobiology, and museology. She has been a student of Great Basin Indian lifeways for thirty years, with studies among Northern Paiute, Southern Paiute, and Timbisha Shoshone peoples. Fowler has worked on several tribal projects in language and cultural maintenance, and the assessment of traditional cultural properties. She is the associate editor of Volume 11 of the Smithsonian Institution's *Handbook of North American Indians,* on the Great Basin, and has most recently published *In the Shadow of Fox Peak: An Ethnography of the Cattail-Eater Northern Paiute People of Stillwater Marsh*. She is currently the president of the Society of Ethnobiology.

Bruce E. Hilpert is the curator of public programs at the Arizona State Museum. An Arizonan since 1965, Hilpert has worked in museum exhibits and education for the past eighteen years. As chief interpreter on the Paths of Life exhibit project, he administers the research, planning, writing, and design process. He also played a leading role in the exhibit design process and wrote the texts of the exhibit's labels. Hilpert has worked at the Arizona Historical Society and has written several articles relating to Arizona history and museum exhibits.

Thomas M. Kolaz is the assistant curator of the ethnological collections at the Arizona State Museum. In addition to fulfilling his curatorial duties, he works with Native American tribes on issues related to the Native American Graves Protection and Repatriation Act. He has actively documented Yaqui mask carvers in the Tucson area and has attended Yaqui fiestas for more than fifteen years.

Nancy J. Parezo is the curator of ethnology at the Arizona State Museum. Dr. Parezo specializes in the art, economics, and material culture of the native peoples of the Southwest. Her best-known work in this area is *Navajo Sandpaintings: From Religious Act to Commercial Art*. She has also worked extensively on the history of anthropology, especially museums, Indian-Anglo interactions, and women anthropologists who have worked in the Southwest. Her work has resulted in several publications, including *Daughters of the Desert* (with Barbara A. Babcock) and *Hidden Scholars*. She has been a fellow of the School of American Research, the Smithsonian Institution, and most recently the D'Arcy McNickle Center of the Newberry Library, where she studied the use of native dress as a visual cue in stereotyping behavior.

Thomas E. Sheridan is the curator of ethnohistory at the Arizona State Museum. Dr. Sheridan specializes in the ethnology, ethnohistory, and political ecology of the southwestern United States and northern Mexico. He served as the director of the Mexican Heritage Project at the Arizona Historical Society, where he wrote *Los Tucsonenses: The Mexican Community in Tucson, 1854–1941* (University of Arizona Press, 1986). He has written or edited seven other books and monographs, including *Where the Dove Calls: The Political Ecology of a Peasant Corporate Community in Northwestern Mexico* (University of Arizona Press, 1988). His most recent work is *Arizona: A History* (University of Arizona Press, 1995). A board member of the Southwestern Mission Research Center, he also edits the quarterly *SMRC Newsletter.*

PLATE AND FIGURE CREDITS

Plates

1 Shell and turquoise necklaces (ASM Neg. C-29873). Left, ASM E-81, length: 40.5 cm. Center, ASM E-9365, length: 57 cm. Right, ASM 21968, length: 47.0 cm.

2 Navajo woman's dress (ASM Neg. C-9766). ASM E-2857, length: 126 cm.

3 Navajo squash blossom necklaces (ASM Neg. C-29872). Left, ASM E-7430, length: 37 cm. Center, ASM E-7428, length: 43 cm. Right, ASM E-1189, length: 43 cm.

4 Navajo Ganado-style rug (ASM Neg. C-29461). ASM 8419, length: 211 cm.

5 Navajo Wide Ruins rug (ASM Neg. C-29463). ASM E-2506, length: 175 cm.

6 Yaqui *sea ania* mural by Danny Leon (ASM Neg. C-29336). Length: 150 cm; width: 242 cm.

7 Mayo *pahkora* masks (ASM Neg. C-29806). Center, ASM 80-13-95, length: 55 cm. Left, ASM 79-135-1096, length: 19.5 cm. Right, ASM 83-3-95, length: 30 cm.

8 Yaqui *pahkola* masks and rattle (*senasum*) (ASM Neg. C-29807). Rattle, ASM E-526, 25.4 cm. Mask at right, ASM 3293, length: 50 cm. Mask at left, ASM E-1254, length: 39.3 cm.

9 *Apache Creation Story* painting (ASM Neg. C-22837). ASM 88-28-1, length: 103 cm.

10 Western Apache coiled basketry and materials (ASM Neg. C-29861). Jar, ASM E-2815, height: 17 cm. Bowl, ASM 23394, diameter: 51 cm. Sumac, ASM E-52.

11 San Carlos Apache fiddle (ASM Neg. C-29863). ASM 84-81-1, length: 63.5 cm.

12 Western Apache bilobe canteen (ASM Neg. C-29859). ASM E-9, height: 33.5 cm.

13 Western Apache painted twined burden basket (ASM Neg. C-29860). ASM 87-22-181, height: 31 cm.

14 Western Apache beaded buckskin moccasins (ASM Neg. C-29862). ASM E-9248, height: 36 cm.

15 Apache war shield (ASM Neg. C-29866). ASM 87-22-97, diameter: 70 cm (not including feathers).

17 Havasupai basketry (ASM Neg. C-29856). Bowl, ASM 91-57-20, diameter: 41.2 cm. Jar, ASM 91-57-24, diameter: 41.0 cm.

18 Hualapai twined basketry (ASM Neg. C-29855). Jar, ASM 79-100-1, height: 18.5 cm. "Fruit bowl," ASM E-4228, height: 23.0 cm.

19 Yavapai coiled basketry (ASM Neg. C-29858). Left, ASM E-2772, diameter: 23.5 cm. Right, ASM E-9355, diameter: 24.5 cm.

20 Hualapai beadwork (ASM Neg. C-29852). Left, ASM E-10,117, length: 50 cm. Right, ASM E-10,119, length: 21 cm. Bottom, ASM E-10,118, length: 71 cm.

21 Tohono O'odham *wi:gida* figures (ASM Neg. C-29815). Left, ASM 92-90-1, height: 38 cm. Right, ASM 92-90-2, height: 31.3 cm.

24 Tohono O'odham wine-making paraphernalia (ASM Neg. C-29826). Jar, ASM 82-41-2, height: 43.9 cm. Pot ring, ASM E-6068-x-5, diameter: 21.5 cm. Coiled wine basket, ASM 1626-x-2, diameter: 38 cm. Ladle, ASM E-4065, length: 51.5 cm. Gathering stick, ASM E-5999, length: 65.5 cm. Strainer basket, ASM E-96, diameter: 28 cm.

25 O'odham bowl baskets (ASM Neg. C-29827). Top right, ASM E-537, diameter: 24 cm. Lower right, ASM E-4708, diameter: 8.8 cm. Center, ASM 26497-x-10, diameter: 21.2 cm. Top left, ASM 18504, diameter: 16 cm. Lower left, ASM E-8798, diameter: 15.9 cm.

27 O'odham pottery (ASM Neg. C-29811). Upper right, wedding vase, ASM 85-25-3, height: 27.6 cm. Lower right, canteen, ASM E-3884, width: 16 cm. Lower left, bowl, ASM 253, width: 15 cm. Upper left, "friendship bowl," ASM 85-25-2, height: 16.7 cm.

28 Tarahumara cross and rosary (ASM Neg. C-29802). Cross, ASM 77-33-458, height: 103 cm. Rosary, ASM 77-33-359, length: 43 cm.

29 Tarahumara violin (ASM Neg. C-29798). ASM 77-33-179, length: 71 cm.

30 Tarahumara Fariseo headdress and sword (ASM Neg. C-29796). Headdress, ASM 77-33-359, height: 46 cm. Sword, ASM 77-33-361, length: 105 cm.

34 Tarahumara Judas figure (ASM Neg. C-29803). ASM 77-33-468, height: 120 cm.

35 Southern Paiute water bottles (ASM Neg. C-29847). Left, ASM 8050, height: 30.1 cm. Center, ASM 8048, height: 21.5 cm. Right, ASM 8052, height: 23.0 cm.

36 San Juan Paiute baskets (ASM Neg. C-29849). Left, ASM 85-14-2, diameter: 38.8 cm. Right, ASM 85-14-1, diameter: 44.7 cm.

37 Southern Paiute wedding baskets (ASM Neg. C-29850). Left, ASM E-4822, diameter: 34.5 cm. Right, ASM E-9204, diameter: 25.5 cm.

38 Pre-Seri or Seri sherds and figures (ASM Neg. C-29780). Sherds, ASM A-36,768-x-1,x-2. Left figure, ASM 75-41-7, height: 14.0 cm. Center, ASM 75-14-12, 14.2 cm. Right, ASM 75-41-1, height: 16.1 cm.

39 Seri seaweed dolls (ASM Neg. C-29783). Left, ASM E-1060, height: 37.0 cm. Right, ASM E-10159, height: 29.0 cm.

40 Seri carrying pouches (ASM Neg. C-29784). Pelican throat bag, ASM 76-108-4, length: 22.2 cm. Bladder canteen, ASM E-925, length: 14 cm. Stomach canteen, ASM 76-108-2, length: 33.7 cm.

41 Seri necklaces (ASM Neg. C-29781). Top, ASM E-520, length: 36.8 cm. Center, ASM E-8594, length: 31.5 cm. Bottom, ASM E-8599, length: 52 cm.

42 Seri coiled basketry (ASM Neg. C-29788). Bowl, ASM E-8289, diameter: 40 cm. Bowl, ASM 83-29-11, diameter: 41.3 cm. Jar, ASM E-5018, height: 16.0 cm.

43 Seri santos (ASM Neg. C-29800). Left, Janus male figure, ASM E-952, height: 8.4 cm. Center, male figure, ASM E-945, height: 13 cm. Pendant, ASM E-819, height: 9 cm.

44 Mohave ceramic figurine pair (ASM Neg. C-29835). ASM 82-55-1, height: 18.0 cm.

47 Colorado Indian Tribe crafts (ASM Neg. C-29844). Kachina, ASM 89-7-1, height: 17.1 cm. Mohave ceramic jar, ASM 76-102-1, height: 22.0 cm. Chemehuevi basketry jar, ASM E-2551, height: 10.5 cm. Navajo rug, ASM 89-7-3, length: 50 cm.

48 Colorado River Yuman tourist crafts (ASM Neg. C-29840). Mohave toad effigy, ASM E-4905, height: 11.4 cm. Quechan flute, ASM 580, length: 42.5 cm. Mohave necklace, ASM 22139, length: 68 cm.

49 Maricopa pottery (ASM Neg. C-29843). Left, jar, ASM 22757, height: 23.5 cm. Center, effigy jar, ASM 22789, height: 27.5 cm. Right, effigy jar, ASM 22879, height 14.5 cm.

50 Hopi Kachina motif crafts (ASM Neg. C-29881). Coiled basketry plaque, ASM E-532, diameter: 29.5 cm. Crow Mother kachina, ASM E-6152, height: 30.0 cm. Tile, ASM 75-13-8, height: 14.6 cm. Wicker plaque, ASM 1921, diameter: 29.0 cm.

51 Hopi woven textiles (ASM Neg. C-29877). Belt, ASM 80, length: 239.5 cm. Brocaded sash, ASM 5437, length: 27.5 cm.

52 Hopi *tableta* (ASM Neg. C-29875). ASM E-6391, height: 44.2 cm.

53 Hano polychrome jar (ASM Neg. C-29887). ASM GP52534, diameter: 37.5 cm.

54 Hopi tiles (ASM Neg. C-29885). Left, ASM 75-13-10, length: 21.6 cm. Right, ASM 8343, length: 17 cm.

55 Hopi corn motif jar (ASM Neg. C-29885). ASM E-9690, height: 17.9 cm.

56 Niman kachina (ASM Neg. C-29878). ASM E-4112, height: 43.0 cm.

Figures

Tohono O'odham calendar stick (ASM Neg. 16779). ASM E-151, length: 168 cm.

Tohono O'odham carrying basket (ASM Neg. 87576). ASM E-2455, diameter: 57 cm.

Tarahumara *tesgüino* paraphernalia (ASM Neg. 87568). Jar, ASM 77-33-8, height: 63.0 cm. Spoon, ASM 77-33-311, length: 67.0 cm. Ceramic bowl, ASM 77-33-156, diameter: 46.0 cm. Gourd

bowl, ASM 77-33-93, diameter: 17.6 cm. Strainer, ASM
77-33-569, diameter: 40.0 cm.

Seri harpoon points (ASM Neg. 87550). Left, ASM E-981, length:
23.5 cm. Center, ASM E-5235, length: 13.0 cm. Right, ASM
E-5863, length: 16.6 cm.

Mohave traveler's kit (ASM Neg. 56611). ASM E-4971, length:
120 cm (bow).

Abee, Juan Isidro Fernández de, 150–51
Agriculture, 232, 240; Diné, 6, 8, 29; Hopi, 239
(fig.), 242, 244, 245, 254, 262–63; Indé, 75–76,
77, 78 (fig.), 80 (fig.); Nïngwï, 168–69;
O'odham, 120–23, 128, 130, 133–36, 138–39;
Rarámuri, 143, 145, 150; Upland Yuman, 94,
103, 107; in Yaqui River Valley, 56, 58
Ak-cin, 124
Akimel O'odham (River People), 110, 116–17,
118, 125, 132 (fig.), 224; agriculture, 126,
133–36, 138–39; basketry, pl. 25; diabetes in,
130, 131; pottery, pl. 27; water rights, xxxiii,
136–39; women, pl. 26
Alarcón, Hernando de, 217, 219–20
Altar River, 133
Anglo Americans, 70, 132 (fig.), 175, 214; and
Diné, 7, 11, 20; and Hopis, 238, 241, 255–57,
265 ; and Nïngwï, 176–77, 179; and Yumans,
91, 100–104, 108, 110, 111, 226–28
animals, 64, 65, 76, 89, 118, 167, 173, 174, 192;
in mythology, 4, 65, 117, 119, 122, 164–65,
175, 215, 216, 240, 241. See also deer; livestock; sheep; turtles, green sea
Anza, Juan Bautista de, 203, 220
Apaches. See Indé
Apache scouts, 67 (fig.), 70, 71, 72
Apaches de Nabajú. See Diné
archaeology, 82–83, 132–33, 143, 242–43
Arizona Volunteers, 230 (fig.)
armies, 10; Mexican, 35, 43, 161, 189, 226; Spanish, 200, 203; U.S., 11–12, 67 (fig.), 68, 70–71,
72, 83, 102–3, 104, 134, 176, 213, 226, 227–
28, 255; Yuman, 224–25, 230 (fig.)
Armistead, Lewis, 228
arts and crafts, 83, 159. See also basketry; blankets; pottery; weaving
Athapaskans, 7, 61–62, 66. See also Diné; Indé

Awatovi, 246, 250, 251, 252
Azul, Antonio, 134–35

Bacatete Mountains (Sierra del Yaqui), 43, 55
Bacavi, 258, 260
Bacum (Vahkom), 35, 43
Bahía Kino, 189 (fig.), 196, 208, 210
Baja California, 187, 188, 189, 218
Baldy, Mount (Dzil Ligai), 61, 62
Banderas, Juan (Juan Ignacio Jusacamea), 42
bands: Comcáac, 196–97; Indé, 66–73; Nïngwï,
181–82, 184; Yuman, 93–94, 219–20
Barboncito, 14
Barranca del Cobre, 143, 145
basketry, 95, 123 (fig.), 159, 248; Chemehuavi,
pl. 47; Havasupai, pl. 17; Hopi, pl. 50;
Hualapai, pl. 18; Nïngwï, 165, 170–71, 183;
O'odham, pl. 24, 25; San Juan Paiute, 35; Seri,
pl. 42; Southern Paiute, pl. 37; Western
Apache, pls. 10, 12, 13; Yavapai, 96–97, pl. 19
Beale, Edward, 101
Bedonkohes, 70
Belem (Veenem), 35
Benavides, Alonso, 9
blankets, 9–10, 20, 221
Bohoroqui, Damian, 41
Borlaug, Norman, 56
Bosque Redondo, 12–13
Buatachive (Son.), 43
Bureau of Indian Affairs (BIA), xxxi (fig.), 58,
128; and Hopi, 257, 259–60, 261, 262, 264–65;
livestock reduction, 23, 26–28; and Nïngwï,
178, 180; and Yumans, 103, 108–9, 110, 229–
30, 231

Cabeza de Vaca, Alvar Nuñez, 198
Cabora, Teresa de (Teresa Urrea), 41

Cajeme (José María Leyva), 42, 43
California, 134, 176, 214, 218, 220, 226–27
California, Gulf of, 187, 244; and Comcáac,
    189–93
Camp Verde, 101, 107, 108
canteen, pl. 12
Canyon Creek, pl. 16
Captain George, 166–67, 173
Cárdenas, Lázaro, 55, 56
Carleton, James, 11–12, 14
Carson, Christopher "Kit," 12
Casa Grande (Ariz.), 117, 125
Casanova, Lola, 207
Casas Grandes (Chih.), 148, 192
Catholicism, 9, 47, 54, 142, 156–57
Central Arizona Project (CAP), 113, 129
Ceremonies, 183; Bear Dance, 168, 172; Cry,
    185; Holy Week celebrations, ii, 47–49, 55,
    142, 157–58, 159, 160, pls. 31–34; Hopi, 243,
    260, 263–65; Indé, 62, 64–65, 81–82; killing-
    the-deer ceremony (lutu pahko), 49; Navajo
    Blessingway, 9, 183; Nïngwï, 173, 185; pascola
    dancers (pahkolam, pahkoram), 36, 41, 47, 49,
    50, 55, pl. 8; Passion plays, 157; Rarámuri,
    156–58, 159, 160, pls. 28, 31–34; Round
    Dance (Nïngwï), 172, 175; Soyala ceremonies,
    260; Tohono O'odham, 119–20, 124–25; Sun-
    rise Ceremony, 64; Yoemem, ii, 46–49, 51–55
Cerro Prieto (Son.), 202, 203
Chacón, Fernando, 9–10
Changing Woman, 4, 5, 64
chapayekam (čhapayékam), 41, 55
Chato, 71, 72
Chemehuevis (Black Bearded One; Southern
    Men), 163, 164, 216, 220, 224, 229, 231, 232;
    basketry of, 170, 171; and Spaniards, 174–75
Cherum, 98, 102–3
Chihuahua, 70, 118, 143
Chiricahua Apache, 66, 70–71, 72
Chona (Maria Encarnacion), 130
Chooksa homar, 214, 217
Christianity: Baptists, 256; conversion to, 151–
    52, 177, 178 (fig.); Easter, 142; Franciscans,
    155, 250, 252; and Hopi, 250–51, 256, 260;
    Jesuits, 36, 40, 46–47, 147, 148, 155, 160,
    198–202; Lent, 47, 158; missionaries, mis-
    sions, xxx, 6, 36, 40, 46–47, 115, 145, 147–48,
    149–51, 154–55, 160, 168, 169, 172, 198–202,
    250–51, 252, 256; Moravians, 256; and
    Rarámuri, 141, 142, 149–51, 160; Virgin
    Mary, 141, 157; and Yoemem, 36, 54
Chu-ar-um-peak, 169 (fig.), 178
Churro, Victoriano, 144

Cibecue Apache, 66, 68, 72, 80–81
Civil War, 11, 134, 230 (fig.)
clans, 251; Diné, 5, 7–8, 10; Hopi, 117, 238–39,
    242–43, 246, 247, 257–58, 263–64
Clarkdale, 108
clothing, 13–14, pls. 2, 14
Cochimí, 188
Cochise, 70
Cocopas, 188, 213, 214, 215, 217, 218, 220, 222,
    223 (fig.), 224, 227, 233, 235
Collier, John, 23, 26, 28, 260
Colorado River, 91, 94, 99, 216; Nïngwï on, 163,
    165; Yumans on, 213, 214, 217–18, 220, 225
Colorado River Basin Project Construction Act,
    113
Colorado River Indian Reservation, 101, 103,
    228–31, 234
Colosio, María Antonia, 190–91
Comanches, 10, 13, 240
Comcáac, 206, 209–11; basketry, pl. 42; cam-
    paigns against, 202–3; desert and, 193–96;
    Europeans and, 197–202; ironwood carvings,
    204–5; necklaces, pl. 41; origins of, 187–89;
    perceptions of, 203, 207; pottery figurines, pl.
    38; pouches, pl. 40; sea and, 189–92; seaweed
    dolls, pl. 39; territory of, 196–97
Concepción, Río (Son.), 115, 133
Coninas, 93
Coolidge Dam, 136, 137 (fig.)
Coronado, Francisco Vásquez de, 66–67
Couts, Cave J., 227
Coyote Iguana, 203
creation, xxx, 125, 142, 192; Diné, 3–6; Hopi,
    238, 240–41; Hualapai, 92–93; Indé, 62–65;
    Mohave, 214, 216–17
Crook, George, 67 (fig.), 70–71, 104
Crozier, Kate, 101

dancers, dances, 41, 100, 256; Hualapais, 106–7;
    Indé, 62, 64; Nïngwï, 168, 172, 184; Tohono
    O'odham, 124, 139; Yoemem, xxvii (fig.), 36,
    48–49, 51–52, 55
Dawaventsie, 254, 255
deer, 39, 76–77; in Yoemem ceremony, 41, 48–
    49, 51–53
deer ceremonies: dancers (maso), xxvii (fig.), 36,
    41, 47, 48–49, 51–52; singers, xxvii (fig.), 41;
    songs (maso bwikam), xxviii, 39, 40, 49, 55, 59
deities, 20, 96, 116, 117, 130, 164, 238, 255;
    Changing Woman, 4, 5, 64; Child Born of
    Water, Child of the Water, 4, 64; Christianiza-
    tion and, 36, 41, 47, 141, 142, 156, 157; in cre-
    ation stories, 4, 5, 62, 64, 65, 91–92, 141, 142,

157, 214, 238, 240, 241; Kachinas, 242, 243, 253, 256, 262, pls. 47, 50, 56, 57; Sun, 62, 142, 157; Talking God, 5. *See also* Devil; Gaan

Desemboque, 192, 210

Desert Indian Dancers, 139

Devil, Rarámuri views of, 141, 142, 157, 158

diabetes, 130–31

Díaz, Porfirio, 45, 160

Diné, xxviii, xxxiii, 32–33, 66, 100, 104, 163, 232, 240, 254, 256; and Colorado River Indian Reservation, 230, 231; creation story of, 3–6; education of, 30–31; and Hopi, 251, 252, 262; jewelry of, pls. 1, 3; livestock of, xxxii (fig.), 8–9, 16 (fig.) 17–19, 22–28; and Long Walk, 11–14; migrations of, 6–8; and Nïngwï, 175, 176, 177, 180, 183, 184; as pastoralists, 3, 4 (fig.), 6, 8–9, 20; treaties of, 14–15; weaving of, 9–10, 20–21, 22–23, pls. 2, 4, 5; World War II and, 28–29. *See also* Navajo Nation

Dinétah, 7, 10

Dodge, Chee, 27

Domínguez, Atanasio, 168, 175

dreams, dreaming, 215, 217

Eastern Pueblos, 246; Jemez, 10; Sandia, 252; Tiwa, 253

education, 29, 30–31, 108–10, 160, 257, 264

Eight Sacred Pueblos, xxviii, 35, 40, 55

ejidos, 41, 160, 161

Encinas, Pascual, 202 (fig.), 207

Escalante, Silvestre Vélez de, 175, 253

Espéjo, Antonio de, 67, 93

families, xxix, 95 (fig.), 185; Diné, 22–23, 24–25; Hopi, 246–47, 248; Indé, 75 (fig.), 77

famines, 26, 255

Fariseos. *See* Pharisees

Fernández de Rentana, Juan, 149, 199

Ferrocarril Chihuahua al Pacifico, 159

fiddle, pl. 11

figures, figurines, 206, pls. 21, 34, 38, 39, 43, 44, 49

First Mesa, 251, 253–54, 261

fishing, by Comcáac, 191, 192, 208, 210

flowers, in Yoemem culture, 39, 51–52

flower world (*sea ania*), 39, pl. 6

Font, Pedro, 221, 225–26

Fonte, Juan, 146, 147

Fort Apache (Ariz.), 69–70

Fort Apache Reservation, 72, 85, 88

Fort Apache Timber Company (FATCO), 89

Fort Fauntleroy Massacre, 11

Fort Marion (Fla.), 71

Fort McDowell Reservation, 107, 108, 110, 113

Fort Sill (Okla.), 70, 71

Fort Sumner, 12–13, 14

Gaan (Mountain Spirit People), 62–63, 64–65

Gabriel, Saint, 36

Gadsden Purchase, 226

games, 49, 144, 233; peón, 233; races, by Rarámuri, 144; running, 130, 144

Garcés, Francisco, 93, 174–75, 218, 221, 225

García Morales, Jesús, 42–43

Gensler, C. H., 229–30

Ge Oidag, 124, 126

Geronimo, 10, 70–71, 72

Gila River, 70, 99, 117, 126, 218, 225; Akimel O'odham and, xxxiii, 115, 133, 134, 135, 138–39

Gila River Farmers Association, 139

Gila River Indian Community, 134

Gilg, Adam, 198, 199

gold rush, 134, 214, 226–27

Grand Canyon, 12, 103, 105, 163

Green Revolution, 56

Ground Mist Girl, 5

Guadalupe (Ariz.), xxviii, 58, 59

Guaymas, 192, 197, 203, 207

Guaymas (Comcáac), 196–97

Gu Oidak, 139–40

haciendas, 43, 160

Halchidhomas, 99, 218–19, 220, 222, 224, 225

Halyikwamais, 218, 219–20, 225

Hamblin, Jacob, 177

Hano (Tewa Village), 253–54, 257

Havasupais (People of the Blue Green Water), 10, 95, 103, 105, 109, 164, 218, 224, 251, 252, 256; basketry of, 97, 170, 171, pl. 17; and Hualapais, 93–94; land of, 91, 92, 104, 105, 112; and Hopis, 99–100, 254

Havasupai Tribal Council, 112

headmen, 9, 172–73, 178

health and healing: disease, 9, 103, 108, 148, 149, 174, 218; healers and healing, 36, 41, 62, 173–74, 253; health care, 32, 182; medicine men, 206; Shamans, 149, 150, 152, 173–74

Heintzelman, Samuel, 227

Hermosillo, 203, 207

Hesperus Peak (Obsidian Mountain), 8

Hiacid O'odham, 224

Hímeri O'odham, 115

Hoffman, Joseph, 73, 77

Hohokam, 116, 117, 118, 124–25, 132, 192

Holy People (Diné), 4, 5, 6, 7, 20

Homol'ovi, 242, 244–45

Hopi Mesas, xxviii, 239, 242–43, 245, 252–53, 262

Hopi-Navajo land dispute, 184, 262

Hopi Reservation, 256–57

Hopis (Hopituh Shi-nu-mu), xxviii, xxxiii, 117, 231, 232, 242; basketry, pl. 50; clans of, 257–58; conflicts in, 258–59; identity of, 237–38, 262–66; joint-use area, 32–33; kachina dolls, pls. 50, 56, 57; origins of, 238–42; and Nɨngwɨ, 163, 168; pottery jar, pl. 55; pottery tiles, pls. 50, 54; and Spaniards, 246–47, 250–53; *tableta*, pl. 52; and U.S. government, 259–62; and Upland Yumans, 97, 98, 99–100; weaving, pl. 51; and whites, 253–57

Hopivotskwani (Hopi Path of Life), 237–38, 240, 241, 242–43, 258, 264

Horcasitas, San Miguel de, 200

Hotevilla, 258, 259–60

housing, 6, 9, 77, 95, 97, 168, 232 (fig.), 246–47, 263 (fig.); caves, 62, 63, 141, 146; tipis, 168

Hualapai Bilingual/Bicultural Education Program, 110, 111

Hualapais (Pine Tree People), xxxi (fig.), 100, 101, 111, 114, 164, 177, 216, 218, 224, 229; basketry of, 97, 170, 171, pl. 18; beadwork, pl. 20; creation of, 92–93; land of, 91, 105, 112; pottery of, 94–95; subgroups of, 93–94; as traders, 98, 222; warfare of, 102–3

Hubbell, Lorenzo, 19, 260

Hubbell Trading Post, 19 (fig.), 20

hunting and gathering: Comcáac, 187, 193–96; by Nɨngwɨ, 165–69, 172; Rarámuri, 143, 145; by Upland Yumans, 94–95; by Indés, 73–75, 76–77, 79

Indé, 5, 7, 61, 70, 103–4, 115, 134, 223 (fig.), 224, 240, 254; bands of, 66–73; basketry of, pls. 10, 12, 13; creation stories, 62–65; fiddle, pl. 11; hunting and gathering by, 73–79; land use, 86–90; moccasins, pl. 14; and mountains, 79–83; shield, pl. 15; stories of, 83–86. *See also various subgroups*

Indian Reorganization Act (IRA), 27–28, 260

Institute of American Indian Art, 64

IRA. *See* Indian Reorganization Act

ironwood (*Olneya tesota*), 204–5

Izábal, Rafael, 44, 45

Jalisco, 117, 118

jewelry, pls. 1, 3, 20, 41, 48

Jicarilla Apache, 66

Jusacamea, Juan Ignacio (Juan Banderas), 40, 42

Kabayum, 47, 55

Kahwan, 218, 219

Kaibab Nɨngwɨ (Mountain Lying Down People), 163, 168, 173–74, 179, 180, 181 (fig.), 185

Kaibab Paiute Tribe, 185

Kamias, 216, 218, 219, 224

Kavelchadom, 218, 219, 224

Kawaikuh, 246, 250

Keam, Thomas, 19

Keams Canyon, 21 (fig.), 255, 256, 257

Kewevkapaya, 94

Kiliwa, 218

Kingman, xxxi (fig.), 109

Kino, Eusebio Francisco, 115, 133, 225

Kiowa-Apache, 66

Kykotsmovi (New Oraibi), 258, 260, 261

labor, Rarámuri as, 148–49, 160

Lagunas, 246

Lamanites, 177, 256

land, 251; Hopi, 237–38, 247, 248, 252, 256–57; Indé, 61, 83–84, 86–90; Nɨngwɨ, 174, 175, 184; rights to, xxxiii, 110, 160, 184; Yuma, 91–93, 100–102, 104, 105, 112–14, 229–31; Yoemem, 43–44, 55–56

Land Allotment Act, 257

languages, xxiii, 41, 110, 111, 112, 117–18, 143, 188

Lehi family, 183

Lehi, Mabel, 166 (fig.)

Lehi, Marie, 171

Leon, Danny, painting by, pl. 6

Leroux, Antoine, 100–101

Lewis, Benson, 83, 84

Lipan Apache, 66

livestock, 133, 141, 150, 197–98, 257; and Akimel O'odham, 133–34; Diné, 6, 8, 17–18, 19, 22, 26; overgrazing by, 17, 32, 33. *See also* horses; ranching; sheep

Loco, 70

logging, 72, 89, 160, 161

Lololoma, 257, 259

Lomahongyoma, 257, 259

Lomawaima, Hartman, 238

Long Walk, 11–14

Lopez, Danny, 139–40

Lower Pimas. *See* Pima Bajos

Luna, Manuel, 144

Lupe, Ronnie, 86

Magdalena (Son.), 218

Magdalena River, 115, 133

Maldonado, Juan (Tetabiate), 43

Mangas Coloradas, 70

March of Tears (Yavapai), 104

Maricopas, 132 (fig.), 134, 135, 214, 215, 217, 235; and Halchidhomas, 218–19; pottery, pl. 49; warfare and, 222, 223–24

Marshall, John, 30

masks, 50, pls. 7, 8

Mastamho, 214, 216–17, 222

matachines (*matachinim, matačinim*), ii, 41, 55

Matavilya, 214

Matcita, 242

Mayos. *See* Yoremem

Mennonites, 256

Mescalero Apache Reservation, 70, 71

Mescalero Apaches, 12, 66

Mexicans, 5, 36, 70, 115, 141, 214, 226, 254

Mexican War, 134, 226

Mexico, xxiv, 70, 99; and Comcáac, 192, 203; and Rarámuri, 160–61; and Yoemem, 40, 42, 55–56

Mexico, government of: Instituto Nacional Indigenista (INI), 160; National Indian Institute, 160; Reform Laws, 160

Michael, Saint, 36

Michler, Nathaniel, 213–14, 219, 225

Michoacán, 191, 192

migrations: Diné, 6–8; Hopi, 240–41

Miles, Nelson, 70, 71

mining, 70; and Diné, xxxiii, 29; and Rarámuri, 147–49, 160, 161; and Yavapais, 101, 107–8

Mishongnovi, 246, 250, 251, 256

Mitchell, Grace Jimulla, 96–97

Moapa Reservation, 177–78

Moapa River, 163, 174

Moccasin, 180

moccasins, pl. 14

Moccasin Spring Kaibabs, 179

Mogollon culture, 82–83

Mohaves, 92, 98, 99, 164, 170, 188, 215, 218, 219, 220, 221, 222, 229, 231, 232, 235; creation story of, 214, 216–17; figures, pl. 44; and Ningwɨ, 168, 177; tourist crafts, pl. 48; warfare, 223–24, 227, 228, pl. 45

Mohave-Apaches. *See* Tonto Apaches; Yavapais

Molina, Felipe, 37, 56; *Yaqui Deer Songs/Maso Bwikam*, 53

Montezuma, Carlos, 109–10

Mormons, 98, 104, 177, 178 (fig.), 256

Mountain Pimas, 118

mountains: sacred, xxviii, 8, 61, 63; Indé and, 73–86, 90

Myths, xxx, 142, 192; animals in, 122, 164–65, 175, 215 ; Diné, 3–6; Hopi, 238, 240–41;

Hualapai, 92–93; Indé, 62–65; Mohave, 214, 216–17; O'odham, 116–17, 118–19, 125; Worlds, Diné, 3–4; Yoeme, 35–38, 40

Naiche, 70, 71

Nampeyo, 254, pl. 53

Nana, 70

Narbona, 11

Native Seeds/SEARCH, 131

Navajo Code Talkers, 28–29

Navajo Community College, 30

Navajo Mountain, 180, 184, 239, 242

Navajo Nation, 29–30, 184, 185

Navajo Reservation, 180

Navajos. *See* Diné

Navɨju, 124, 125

New Pascua Pueblo, 58

Ningwɨ, 10, 32, 94, 98, 163; basketry, pl. 35; and Coyote, 164–65; customs, 184–85; hunting and gathering by, 165–69; social organization of, 172–73; and Spaniards, 174–76; and U.S. government, 177–82

Northern Tonto Apache, 66, 68, 72

Obregón, Alvaro, 46

ocean, Comcáac use of, 189–93

O'dami, 118

Oñate, Juan de, 217–18, 250

O'odham, xxx, 115, 116–18, 130–31, 202; Rain, 119, 121. *See also* Akimel O'odham; Tohono O'odham

Opatas, 42, 144

Oraibi, 242, 246, 250, 256, 257, 261; split of, 258–60

Orme Dam, 113

overgrazing, 17, 32–33

Pa'a. *See* Pais

pahkola masks (Yaqui), pl. 8

pahkora masks (Mayo), pl. 7

Painter, Maria, 170

Pais, 72, 91–92. *See also* Havasupais; Hualapais; Yavapais

Paiutes, 10, 99, 242, 256. *See also* Ningwɨ

Paiute Strip Reservation, 180

Paiute Tribe of Utah, 182

Pascua, 58, 59

Pascua Yaqui Association, 58

Peabody Coal Company, 262

Peach Springs, 101, 105

Perera, Nicolás, 199–200

Pérez de Ribas, Andrés, 193, 298

Perry, Edgar, 61, 62

Pharisees (*fariseos, pariséo, pariseram*), 41, 47, 55, 157, 158, pls. 30, 31, 32
Pima Altas, 115. *See also* O'odham
Pima Bajos, 115, 116, 118, 154, 197, 202
Pimas, 42, 98, 164. *See also* Akimel O'odham
Place Where the Waters Crossed, The, 3
Place Where the White Rock Stands, 5
plants, 118, 165; Comáac use of, 192–95, 204, 206; Indé use of, 73–76, 77; O'odham use of, 116, 130–31, 133–34; Old World, 133–34. *See also various types*
Plateau People (Ko'audva Kopaya), 93
Polacca, 255, 257
Pópulo, Santa María del, 199–200
Potam (Son.), 35, 54 (fig.), 58
pottery, 95, 159, 174, 188, 195, 242, 248, 254, pls. 27, 47, 49, 53, 55
pouches, pl. 40
poverty, 26, 32
Powell, John Wesley, 169, 170, 178–79, 181 (fig.)
prehistoric sites, 82–83, 132–33, 143, 148; and Hopis, 238–39, 242
Price, William, 102
property ownership, 247, 248
Pueblo Indians, 246, 251, 252–53; and Diné, 5, 6, 8, 9, 10; and Tohono O'odham, 124, 125
Pueblo Revolt, 10, 251
Punta de Agua, 126–27

Quechan, 99, 164, 188, 213, 214, 215, 216, 217, 218, 219, 220, 225, 229, 235, pl. 46; and gold rush, 226–27; tourist crafts, pl. 48; warfare of, 222–24
Querechos, 67
Quitovac, 124

Rahum, 35, 58
raiding, 254, 255; Comcáac, 188, 203; and Diné, 10, 11, 252; Indé, 70, 134; Yuman, 100–101, 222, 224, 225–26
railroads, 70, 112, 159, 160
ranching, 70, 72, 87–88, 101, 105, 180
Rarámuri, ii, 118, 141, 142, 144, 159, 160–61; cross and fiddle, pl. 28; drinking parties of, 152–55; missionaries and, 147–50; pueblos, 155–56; religion, 151–52, 156–58, pls. 28, 31–34; society of, 145–47; transhumance of, 143, 150–51; violin, pl. 29
rebellions, 203; Rarámuri, 147–48, 149–50, 160; Yoemem, 40, 42–43
recreation, and White Mountain Apaches, 72, 88–89

Religion, 41, 65; Apache, pl. 9; Hopi, 242, 248, 257–58, 262, 263–65, , pls. 52, 54; Hualapais, 106–7; Mayo, pl. 6; Ningwɨ, 173, 185; Rarámuri, ii, 141, 142, 149–50, 156–58, 160; Seri, pl. 43; Snake Society, 255; Tarahumara, pl. 28–34; Tohono O'odham, pl. 21; visions, 205, 215; Yaqui, pls. 6, 8; Yoemem, xxviii, 36, 45, 46–49, 51–53. *See also* Catholicism; Christianity
relocation (removal), 231; Diné, 12–14, 33, 230; Upland Yuman, 103–4; Indé, 71, 72
repartimiento system, 148, 161. *See also* labor
reservations: Apache, 68–69 (fig.); Diné, 15, 17, 28, 29, 32–33; Ningwɨ, 177–78; Tohono O'odham, 132–33; Yuman, 103–8, 110, 112–14, 228–31
Rio Grande Pueblos, 246, 252–53
River Yumans (River Branch), 163, 218. *See also* Cocopas; Maricopas; Mohaves; Quechan
Roessel, Ruth, 30
Romero, Miguel, 54 (fig.)
rugs, Diné, 19, 31, pl. 4, 5, 47

sacred places, xxviii, 8, 61, 63, 175
saguaro, 73, 118, 119–20, 121, 192
St. George (Utah), 179
Salineros, 196, 197
Salt River Indian Reservation, 135
Salt River Pima-Maricopa Reservation, 110
Salt River Valley, 58, 115, 117, 135
San Carlos Apaches, 66, 67–68, 87–88, 107
San Carlos Project, 136, 138
San Carlos Reservation, 68–69 (fig.), 70, 72, 87, 104, 107
San Esteban Island, 188, 189, 190, 192
San Francisco Peaks (Abalone Shell Mountain), 8, 91, 245
San Juan Mountains, 5
San Juan Ningwɨ (Rock River People), 163, 171, 174, 175, 176, 180, 183, 184, 185; basketry, pls. 36, 37
San Juan River, 29
Santa Cruz River, 115, 126–28, 133
santos, Comcáac, 206
San Xavier del Bac, xxx (fig.), 126, 132
San Xavier Development Project, 129
San Xavier district, 129, 132–33
San Xavier Reservation, 126
SAWRSA. *See* Southern Arizona Water Rights Settlement Act
schools, xxxi (fig.), 30, 109, 160, 254, 257, 259, 260
Sekaquaptewa, Emory, 242, 264

Seris. *See* Comcáac

sheep, 150; Diné and, xxviii, xxxii (fig.), 4 (fig.), 6, 8, 16 (fig.), 17–19, 22–28, 31. *See also* livestock

shield, pl. 15

Shivwits (People Who Live to the East), 163, 168, 182

Shivwits Reservation, 179

Shungopavi, 246, 250, 251, 252, 259, 261

Sichomovi, 251

Sierra Azul, 70, 71

Sierra Madre Occidental, 70, 118, 141, 143

Sikyatki, 246, 250

silversmithing, 18 (fig.), 253–54

Sinaloa River, 41

Sine, Duke, *Apache Creation Story*, 64, pl. 9

Sipaulovi, 251

Sitgreaves, Lorenzo, 100–101

Skeleton Cave Massacre, 104

slave trade, 9, 10, 11, 99, 100, 147, 176, 225–26, 254

Soba, 115

Sobaipuri, 115

Soldados (Soldiers), 156, 158

song(s), xxx, 173, 187; Tohono O'odham, 116, 124; Yoemem, 39, 40, 47–48, 49

Sonora, xxviii, 35, 41, 70, 143; Comcáac in, 187–89, 192–96; O'odham in, 115, 118; Yoemem in, 40, 42–43

Sonoran Desert, 73; Comcáac use of, 193–96; O'odham and, 116, 118–19; rain and, 120–21

Southern Arizona Water Rights Settlement Act (SAWARSA), 129, 132

Southern Pacific Railroad, 70, 232

Southern Paiutes. *See* Nɨngwɨ

Southern Tonto Apache, 66, 68, 72

Spaniards, 70, 115; and Comcáac, 189, 196–97, 198–202; and Diné, 7, 9–10; and Hopis, 246–47, 250–53; and Nɨngwɨ, 169, 172, 174–76; and Rarámuri, 143, 147 ; and Yoemem, 36, 40; and Yumans, 93, 100, 217–18, 224, 225–26

Spirit Mountain, 92, 216

Stewart, Irene, 6, 13, 22, 24, 25, 27, 30, 32

stock reduction, 23, 26–28

stories, storytelling, 173, 233; Hopi, 238, 240–41; Indé, 83–86; Mohave, 214, 216–17; Yuman, 96–97

Sunrise Ski Resort, 89

Surem, 37–39

Talayesva, Don, 258

Talking Tree (Yoemem), 36–38, 56

Tarahumara. *See* Rarámuri

Tastioteños, 197

Tepahues, 41

Tepecanos, 118

Tepehuanes (O'dami), 118, 145, 147

Tepiman corridor, 118

tesgüinadas, 152–55

Tetabiate (Juan Maldonado), 43

Tewageh (Tsawadeh), 253

Tewaquaptewa, 259, 260

Tewas, 9, 10, 252, 253–54

Thompson, Nick, 83, 84, 86, 90

Tiburón Island, 187, 188, 189, 195–96, 202, 203, 206, 207, 208

Tohono O'odham (Desert People), 115, 116, 118; agriculture, 120–23; alliances of, 224, 225; basketry, pl. 25; and diabetes, 130–31; pottery, pl. 27; saguaro picking by, pls. 22, 23; saguaro wine making by, 119–20, pl. 24; trade with, 99, 221; water rights of, 126–29; *wi:gida* ceremony, 124–25, pl. 21

Tokonave, 239

Tolkapaya, 94

Tonto Apaches, 72, 107, 224. *See also* Northern Tonto Apaches; Southern Tonto Apaches

tourism, 112, 159, 208 (fig.), 209 (fig.), 210

traders, 19, 98–99, 221–22

Trail of Tears (Yavapai), 113

Treaty of Guadalupe Hidalgo, 226

Trincheras culture, 192

Tsawadeh, 253

Tsegi Canyon, 5, 176

Tuba, Chief, 255–56

Tucson: and water rights, 126–29; Yoemem in, 45, 58

Tucson Basin, water in, 126–29

turtles, green sea (*Chelonia mydas*), 191–93

Uinkarets (People of the Ponderosa Pine Peak), 163–64

U.S. government, xxiv, 30; and Akimel O'odham, 135, 136–38; Bureau of Reclamation, 113; Civilian Conservation Corps (CCC), 27; and Diné, 11–15, 17, 23, 26–28; Emergency Conservation Works program, 26; Federal Emergency Relief Administration, 23; and Hopi, 254, 256–57, 260–262; and Indé, 68, 70, 71, 72; Indian Claims Commission, 101, 105; National Park Service, 112; and Nɨngwɨ, 177–82, 184; Office of Economic Opportunity, 58; Yoemem and, 45, 58; and Yumans, 102–8, 112–14, 213–14, 228–31. *See also* Bureau of Indian Affairs

Upland Yumans, 91, 163, 170, 218. *See also*
    Havasupais; Hualapais; Yavapais; culture of,
    93–95, 98–100; education of, 108–10; land
    rights of, 112–14; reservations, 105–8, 228
Upper Pimas, 115. *See also* O'odham
Utah, 163, 177, 180, 182
Utes, 183, 251, 254; and Diné, 5, 10, 12; and
    Hopi, 240, 242, 252; and Nɨngwɨ, 163, 168,
    176

Vargas, Diego de, 251
Verplanken, Luis, 159
Vicam (Vikam), 35, 38, 58
Victorio, 72
violin, pl. 29
Virgin River, 169, 172, 175
Voth, Henry R., 256
voting rights, 110

Walapai Charlie, 108–9
Walpi, 246, 250, 251, 252, 253–54, 255
wars, warfare, 177; Comcáac, 202, 203, 207;
    Indé, 69, 70–71, 72; Diné, 10–11, 12;
    Rarámuri, 147–48; Yoemem and, 40, 42–43,
    45; Yuman, 102–3, 217, 218, 222–28
Watahomigie, Lucille, 110, 111
water, water rights, 110, 113, 195–96; and
    O'odham, xxxiii, 116, 126–29, 131–32, 134–
    39
Wauba Yuma, 101
weaving, 41, 159; Diné, 9–10, 20–21, 22–23, 27
    (fig.), 31; pls. 2, 4, 5
Webb, George, 136–38
Wescogame, Mary, 101
Western Apaches. *See* Indé; *various subgroups*
Whipple, A. W., 255
White, Elizabeth, Hopi jar by, pl. 55
White Mountain Apache Reservation, 69, 72
White Mountain Apaches, 61, 66, 72, 88
White Mountain Game and Fish Department, 89
White Mountain Recreation Enterprises, 88–89
Whiteriver, 86, 89
wickiups. *See* gowa
wi:gida, 124–25, pl. 21
Wikahme, 92
Williams, 109
Williams, John, 91–92, 101–2, 113
Williams v. Lee, 30
Winslow, 244
Winters Doctrine, 128
Wipukpaya, 94
wool, Diné use of, 9–10, 17–19, 22–23

World War II, 28–29
Wovoka, 106
Wupatki, 239

Xavier, Juan, 124
Xomace'kupa'ra, 215

Yahoya, 242
Yaqui Indigenous Zone, 55, 56
Yaqui River Valley, xxviii, 35, 41, 58, 197; rights
    to, xxxiii, 36, 43–44, 55–56
Yaquis. *See* Yoemem
Yaqui Wars, 43
Yava, Albert, 239
Yavapai-Apaches. *See* Yavapais
Yavapais, 93, 94, 95, 216, 218, 224, 229; and An-
    glo Americans, 101–2, 227; basketry, 96–97,
    pl. 19; education of, 109–10; land of, 91, 92,
    113; relocation of, 103–4, 107–8; on San
    Carlos Reservation, 70, 72, 104; trade, 98–99
Yazzi, Mrs. Y. N., 26
Yoemem, xxvii (fig.), xxviii, 35, 39, 41, 147, 197;
    diaspora, xxvi, 44–45; land rights, xxxiii, 43–
    44, 55–56; mythology, plate, 36–38; Pro-
    grama de Atención Técnica Integral para las
    Comunidades Yaquis (PATICY), 58; rebellion
    by, 40, 42–43; religion of, xxviii, 46–49, 51–
    53; role of ritual for, 53–55; Running of the
    Gloria, 47–48, 55; wilderness world (*huya
    ania*), 38, 47, 48, 50
Yoem Pueblo, xxvii (fig.), 58
Yokioma, 259
Yoremem, 41, 42, 50, plate
Young, Captain, 102–3
Yuma-Apaches. *See* Tonto Apaches; Yavapais
Yumans, 217–18.; reservations for, 228–29; so-
    cial organization of, 219–20; tourist crafts, pl.
    48; and U.S. government, 213–14; warfare of,
    222–28. *See also* Cocopa; Havasupais;
    Hualapais; Maricopas; Mohaves; Quechan;
    Upland Yumans; Yavapais

Zunis, 5, 10, 98, 222, 246, 251, 252, 253, 254